Progress in Mathematics 5

Edited by
J. Coates and
S. Helgason

Sigurdur Helgason

The Radon Transform

Birkhäuser
Boston, Basel, Stuttgart

Author

Sigurdur Helgason

Department of Mathematics
Massachusetts Institute of Technology
Cambridge, MA 02139
U.S.A.

Library of Congress Cataloging in Publication Data

Helgason, Sigurdur, 1927—
The Radon transform.

(Progress in mathematics; 5)
Bibliography: p.
Includes index.
1. Radon transforms. I. Title. II. Series: Progress in mathematics (Cambridge); 5.
QA649.H44 516.3′6 80-15951
ISBN 3-7643-3006-6

CIP—Kurztitelaufnahme der Deutschen Bibliothek

Helgason, Sigurdur:
The radon transform / Sigurdur Helgason.—Boston, Basel, Stuttgart: Birkhäuser, 1980.
(Progress in mathematics : 5)
ISBN 3-7643-3006-6

ISBN 3-7643-3006-6

Printed in USA

TO ARTIE

PREFACE

The title of this booklet refers to a topic in geometric analysis which has its origins in results of Funk [1916] and Radon [1917] determining, respectively, a symmetric function on the two-sphere $\mathbb{S}^2$ from its great circle integrals and a function on the plane $\mathbb{R}^2$ from its line integrals (See References). Recent developments, in particular applications to partial differential equations, X-ray technology, and radioastronomy, have widened interest in the subject.

These notes consist of a revision of lectures given at MIT in the Fall of 1966, based mostly on my papers during 1959 - 1965 on the Radon transform and some of its generalizations. (The term "Radon transform" is adopted from John [1955]). The viewpoint for these generalizations is as follows.

The set of points on $\mathbf{S}^2$ and the set of great circles on $\mathbf{S}^2$ are both homogeneous spaces of the orthogonal group $\mathbb{O}(3)$. Similarly, the set of points in $\mathbb{R}^2$ and the set of lines in $\mathbb{R}^2$ are both homogeneous spaces of the group $M(2)$ of rigid motions of $\mathbb{R}^2$. This motivates our general Radon transform definition from [1965A,B] which forms the framework of Chapter II: Given two homogeneous spaces G/K and G/H of the same group G the Radon transform $f \longrightarrow \hat{f}$ maps functions f on the first space to functions $\hat{f}$ on the second space. For $\xi \in G/H$, $\hat{f}(\xi)$ is defined as the (natural) integral of f over the set of points $x \in G/K$ which are incident to ξ in the sense of Chern [1942]. The problem of inverting $f \longrightarrow \hat{f}$ is worked out in a few cases.

It happens when G/K is a Euclidean space, and more generally when G/K is a Riemannian symmetric space, that the natural differential operators D on G/K are transferred by $f \longrightarrow \hat{f}$ into much more manageable differential operators $\hat{D}$ on G/H; the connection is $(Df)^{\wedge} = \hat{D}\hat{f}$. Then the theory of the transform $f \longrightarrow \hat{f}$ has significant applications to the study of the properties of D.

On the other hand, applications of the original Radon transform on $\mathbb{R}^2$ to X-ray technology and radioastronomy are based on the fact that for an unknown density f, X-ray attenuation measurements give $\hat{f}$ directly and therefore yield f via Radon's inversion formula. More precisely, let B be a convex body, $f(x)$ its density at the point x, and suppose a thin beam of X-rays is directed at B along a line ξ. Then the line integral

$$\hat{f}(\xi) = \int_{\xi} f(x)\, dm(x)$$

equals $\log (I_0/I)$ where I_0 and I, respectively, are the intensities of the beam before hitting B and after leaving B. Thus, while the function f is at first unknown, the function $\hat{f}$ is determined by the X-ray data.

The lecture notes indicated above have been updated a bit by the inclusion of a short account of some applications (Chapter I, §7), by adding a few corollaries (2.8, 2.12, 6.4 in Chapter I, 2.8 and 4.1 in Chapter III), and by giving indications in the bibliographical notes of some recent developments.

An effort has been made to keep the exposition rather elementary. The distribution theory and the theory of Riesz potentials, occasionally needed in Chapter I, is reviewed in some detail in §8.

Apart from the general homogeneous space framework in Chapter II, the treatment is restricted to Euclidean and isotropic spaces (spaces which are "the same in all directions"). For more general symmetric spaces the treatment is postponed (except for §4 in Chapter III) to another occasion since more machinery from the theory of semisimple Lie groups is required.

I am indebted to R. Melrose and R. Seeley for helpful suggestions and to F. Gonzales and J. Orloff for critical reading of parts of the manuscript.

Since J. Radon's pioneering paper (1917) is now somewhat inaccessible, it is reproduced here in an appendix.

CONTENTS

CHAPTER I

THE RADON TRANSFORM ON $\mathbb{R}^n$

§1. Introduction.

It was proved by J. Radon in 1917 that a differentiable function on $\mathbb{R}^3$ can be determined explicitly by means of its integrals over the planes in $\mathbb{R}^3$. Let $J(\omega,p)$ denote the integral of f over the hyperplane $(x,\omega) = p$, ω denoting a unit vector and $(,)$ the inner product. Then

$$f(x) = -\frac{1}{8\pi^2} L_x \left(\int_{\mathbb{S}^2} J(\omega, (\omega,x))\, d\omega \right) ,$$

where L is the Laplacian on $\mathbb{R}^3$ and $d\omega$ the area element on the sphere $\mathbb{S}^2$ (cf. Theorem 3.1).

We observe that the formula above contains two integrations dual to each other: first one integrates over the set of points in a hyperplane, then one integrates over the set of hyperplanes passing through a given point. This suggests considering the transform $f \to \hat{f}$, $\phi \to \check{\phi}$ defined below.

The formula has another interesting feature. For a fixed ω the integrand $x \to J(\omega,(\omega,x))$ is a <u>plane wave</u>, that is a function constant on each plane perpendicular to ω. Ignoring the Laplacian the formula gives a continuous decomposition of f into plane waves. Since a plane wave amounts to a function of just one variable (along the normal to the planes) this decomposition can sometimes reduce a problem for $\mathbb{R}^3$ to a similar problem for $\mathbb{R}$. This principle has been particularly useful in the theory of partial diffential equations.

The analog of the formula above for line integrals is of

importance in radiography where the objective is the description of a density function by means of certain line integrals.

In this chapter we discuss relationships between a function on $\mathbb{R}^n$ and its integrals over k-dimensional planes in $\mathbb{R}^n$. The case $k = n - 1$ will be the one of primary interest.

§2. The Radon Transform of the Spaces $C_c^\infty(\mathbb{R}^n)$ and $\mathcal{S}(\mathbb{R}^n)$. The Support Theorem.

Let f be a function on $\mathbb{R}^n$, integrable on each hyperplane in $\mathbb{R}^n$. Let $\mathbb{P}^n$ denote the space of all hyperplanes in $\mathbb{R}^n$, $\mathbb{P}^n$ being furnished with the obvious topology. The Radon transform of f is defined as the function $\hat{f}$ on $\mathbb{P}^n$ given by

$$\hat{f}(\xi) = \int_\xi f(x)\,dm(x) ,$$

where dm is the Euclidean measure on the hyperplane ξ. Along with the transformation $f \to \hat{f}$ we consider alos the dual transform $\phi \to \check{\phi}$ which to a continuous function ϕ on $\mathbb{P}^n$ associates the function $\check{\phi}$ on $\mathbb{R}^n$ given by

$$\check{\phi}(x) = \int_{x\in\xi} \phi(\xi)\,d\mu(\xi)$$

where $d\mu$ is the measure on the compact set $\{\xi \in \mathbb{P}^n : x \in \xi\}$ which is invariant under the group of rotations around x and for which the measure of the whole set is 1. We shall relate certain function spaces on $\mathbb{R}^n$ and on $\mathbb{P}^n$ by means of the transforms $f \to \hat{f}$, $\phi \to \check{\phi}$; later we obtain explicit inversion formulas.

Each hyperplane $\xi \in \mathbb{P}^n$ can be written $\xi = \{x \in \mathbb{R}^n : (x,\omega) = p\}$ where $(,)$ is the usual inner product, $\omega = (\omega_1,\ldots,\omega_n)$ a unit vector and $p \in \mathbb{R}$. Note that the pairs (ω,p) and $(-\omega,-p)$ give the same ξ; the mapping $(\omega,p) \to \xi$ is a double covering of $\mathbb{S}^{n-1} \times \mathbb{R}$ onto $\mathbb{P}^n$. Thus $\mathbb{P}^n$ has a canonical manifold structure with respect to which this covering map is differentiable and regular. We thus identify continuous (differentiable) functions ϕ on $\mathbb{P}^n$ with continuous (differentiable) functions ϕ on $\mathbb{S}^{n-1} \times \mathbb{R}$ satisfying $\phi(\omega,p) = \phi(-\omega,-p)$. Writing $\hat{f}(\omega,p)$ instead of $\hat{f}(\xi)$ and f_t for the translated function $x \to f(t+x)$ we have

$$\hat{f}_t(\omega,p) = \int_{(x,\omega)=p} f(x+t)\,dm(x) = \int_{(y,\omega)=p+(t,\omega)} f(y)\,dm(y)$$

so

$$\hat{f}_t(\omega,p) = \hat{f}(\omega,p+(t,\omega)). \tag{1}$$

Taking limits we see that if $\partial_i = \partial/\partial x_i$

$$(\partial_i f)^{\wedge}(\omega,p) = \omega_i \frac{\partial \hat{f}}{\partial p}(\omega,p). \tag{2}$$

Let L denote the Laplacian $\sum_i \partial_i^2$ on $\mathbb{R}^n$ and let $\square$ denote the operator

$$\phi(\omega,p) \to \frac{\partial^2}{\partial p^2}\phi(\omega,p),$$

which is a well-defined operator on $C^\infty(\mathbb{P}^n)$. It can be proved that if $M(n)$ is the group of isometries of $\mathbb{R}^n$, then L (respectively $\square$) generates the algebra of $M(n)$-invariant differential operators on $\mathbb{R}^n$ (respectively $\mathbb{P}^n$).

LEMMA 2.1. <u>The transforms</u> $f \to \hat{f}$, $\phi \to \check{\phi}$ <u>intertwine</u> L <u>and</u> $\square$, i.e.,

$$(Lf)^{\wedge} = \square(\hat{f}), \qquad (\square\phi)^{\vee} = L\check{\phi} .$$

Proof. The first relation follows from (2) by iteration. For the second we just note that for a certain constant c

(3) $$\check{\phi}(x) = c\int_{\mathbb{S}^{n-1}} \phi(\omega, (x,\omega))d\omega,$$

where $d\omega$ is the usual measure on $\mathbb{S}^{n-1}$.

The Radon transform is closely connected with the Fourier transform

$$\tilde{f}(u) = \int_{\mathbb{R}^n} f(x)e^{-i(x,u)}dx \qquad u \in \mathbb{R}^n .$$

In fact, if $s \in \mathbb{R}$, ω a unit vector,

$$\tilde{f}(s\omega) = \int_{-\infty}^{\infty} dr \int_{(x,\omega) = r} f(x)e^{-is(x,\omega)}dm(x)$$

so

(4) $$\tilde{f}(s\omega) = \int_{-\infty}^{\infty} \hat{f}(\omega,r)e^{-isr}dr.$$

This means that the n-dimensional Fourier transform is the 1-dimensional Fourier transform of the Radon transform. From (4), or directly, it follows that the Radon transform of the

$$f(x) = \int_{\mathbb{R}^n} f_1(x-y)f_2(y)dy$$

is the convolution

(5) $$\hat{f}(\omega,p) = \int_{\mathbb{R}} \hat{f}_1(\omega,p-q)\hat{f}_2(\omega,q)dq .$$

§2

Although slightly greater generality is possible we shall work with the space $\mathcal{S}(\mathbb{R}^n)$ of complex-valued rapidly decreasing functions on $\mathbb{R}^n$. We recall that $f \in \mathcal{S}(\mathbb{R}^n)$ if and only if for each polynomial P and each integer $m \geq 0$,

$$\sup_x \left| \, |x|^m P(\partial_1, \ldots, \partial_n) f(x) \, \right| < \infty, \tag{6}$$

$|x|$ denoting the norm of x. We now formulate this in a more invariant fashion.

LEMMA 2.2. *A function* $f \in C^\infty(\mathbb{R}^n)$ *belongs to* $\mathcal{S}(\mathbb{R}^n)$ *if and only if for each pair* $k, \ell \in \mathbb{Z}^+$

$$\sup_{x \in \mathbb{R}^n} \left| (1 + |x|)^k (L^\ell f)(x) \right| < \infty.$$

This is easily proved just by using Fourier transforms.

In analogy with $\mathcal{S}(\mathbb{R}^n)$ we define $\mathcal{S}(\mathbf{S}^{n-1} \times \mathbb{R})$ as the space of C^∞ functions ϕ on $\mathbf{S}^{n-1} \times \mathbb{R}$ which for any integers $k, \ell \geq 0$ and any differential operator D on $\mathbf{S}^{n-1}$ satisfy

$$\sup_{\omega \in \mathbf{S}^{n-1}, \, r \in \mathbb{R}} \left| (1 + |r|^k) \frac{d^\ell}{dr^\ell} (D\phi)(\omega, r) \right| < \infty. \tag{7}$$

The space $\mathcal{S}(\mathbf{P}^n)$ is then defined as the set of $\phi \in \mathcal{S}(\mathbf{S}^{n-1} \times \mathbb{R})$ satisfying $\phi(\omega, p) = \phi(-\omega, -p)$.

LEMMA 2.3. *For each* $f \in \mathcal{S}(\mathbb{R}^n)$ *the Radon transform* $\hat{f}(\omega, p)$ *satisfies the following condition: For* $k \in \mathbb{Z}^+$ *the integral*

$$\int_{\mathbb{R}} \hat{f}(\omega, p) p^k \, dp$$

can be written as a kth *degree homogeneous polynomial in* $\omega_1, \ldots, \omega_n$.

Proof. This is immediate from the relation

$$(8) \qquad \int_{\mathbb{R}} \hat{f}(\omega,p)p^k dp = \int_{\mathbb{R}} p^k dp \int_{(x,\omega)=p} f(x)dm(x) = \int_{\mathbb{R}^n} f(x)(x,\omega)^k dx.$$

In accordance with this lemma we define the space

$$\mathcal{S}_H(\mathbb{P}^n) = \left\{ F \in \mathcal{S}(\mathbb{P}^n) : \begin{array}{l} \text{For each } k \in \mathbb{Z}^+, \int_{\mathbb{R}} F(\omega,p)p^k dp \\ \text{is a homogeneous polynomial} \\ \text{in } \omega_1,\dots,\omega_n \text{ of degree } k \end{array} \right\}.$$

With the notation $\mathcal{D}(\mathbb{P}^n) = C_c^\infty(\mathbb{P}^n)$ we write

$$\mathcal{D}_H(\mathbb{P}^n) = \mathcal{S}_H(\mathbb{P}^n) \cap \mathcal{D}(\mathbb{P}^n) .$$

According to Schwartz [1966], p. 249, the Fourier transform $f \to \tilde{f}$ maps the space $\mathcal{S}(\mathbb{R}^n)$ onto itself. We shall now settle the analogous question for the Radon transform.

THEOREM 2.4. (<u>The Schwartz theorem</u>). <u>The Radon transform</u> $f \to \hat{f}$ <u>is a linear one-to-one mapping of</u> $\mathcal{S}(\mathbb{R}^n)$ <u>onto</u> $\mathcal{S}_H(\mathbb{P}^n)$.

<u>Proof</u>. Since

$$\frac{d}{ds}\tilde{f}(s\omega) = \sum_{i=1}^{n} \omega_i(\partial_i\tilde{f})$$

it is clear from (4) that for each fixed ω the function $r \to \hat{f}(\omega,r)$ lies in $\mathcal{S}(\mathbb{R})$. For each $\omega_0 \in \mathbb{S}^{n-1}$ a subset of $\{\omega_1,\dots,\omega_n\}$ will serve as local coordinates on a neighborhood of ω_0 in $\mathbb{S}^{n-1}$. To see that $\hat{f} \in \mathcal{S}(\mathbb{P}^n)$, it therefore suffices to verify (7) for $\phi = \hat{f}$ on an open subset $N \subset \mathbb{S}^{n-1}$ where ω_n is bounded away from 0 and $\omega_1,\dots,\omega_{n-1}$ serve as coordinates, in terms of which D is expressed. Since

$$(9) \qquad u_1 = s\omega_1,\dots, u_{n-1} = s\omega_{n-1}, u_n = s(1-\omega_1^2-\dots-\omega_{n-1}^2)^{\frac{1}{2}}$$

we have

$$\frac{\partial}{\partial\omega_i}(\tilde{f}(s\omega)) = s\frac{\partial\tilde{f}}{\partial u_i} - s\omega_i(1-\omega_1^2-\dots-\omega_{n-1}^2)^{-\frac{1}{2}}\frac{\partial\tilde{f}}{\partial u_n} .$$

It follows that if D is any differential operator on $\mathbb{S}^{n-1}$ and if $k,\ell \in \mathbb{Z}^+$ then

$$\sup_{\omega \in N,\ s \in \mathbb{R}} \left| (1+s^{2k}) \frac{d^\ell}{ds^\ell} (D\tilde{f})(\omega,s) \right| < \infty . \tag{10}$$

We can therefore apply D under the integral sign in the inversion formula to (4),

$$\hat{f}(\omega,r) = \frac{1}{2\pi} \int_{\mathbb{R}} \tilde{f}(s\omega) e^{isr} ds$$

and obtain

$$(1+r^{2k}) \frac{d^\ell}{dr^\ell} \left[D_\omega(\hat{f}(\omega,r)) \right] = \frac{1}{2\pi} \int \left[1+(-1)^k \frac{d^{2k}}{ds^{2k}} \right] \left[(is)^\ell D_\omega(\tilde{f}(s\omega)) \right] e^{isr} ds.$$

Now (10) shows that $\hat{f} \in \mathcal{S}(\mathbb{P}^n)$ so by Lemma 2.3, $\hat{f} \in \mathcal{S}_H(\mathbb{P}^n)$.

Because of (4) and the fact that the Fourier transform is one-to-one it only remains to prove the surjectivity in Theorem 2.4. Let $\phi \in \mathcal{S}_H(\mathbb{P}^n)$. In order to prove $\phi = \hat{f}$ for some $f \in \mathcal{S}(\mathbb{R}^n)$ we put

$$\Phi(s,\omega) = \int_{-\infty}^{\infty} \phi(\omega,r) e^{-irs} dr .$$

Then $\Phi(s,\omega) = \Phi(-s,-\omega)$ and $\Phi(0,\omega)$ is a homogeneous polynomial of degree 0 in $\omega_1,\dots,\omega_n$, hence constant. Thus there exists a function F on $\mathbb{R}^n$ such that

$$F(s\omega) = \int_{\mathbb{R}} \phi(\omega,r) e^{-irs} dr .$$

While F is clearly smooth away from the origin we shall now prove it to be smooth at the origin too; this is where the homogeneity condition in the definition of $\mathcal{S}_H(\mathbb{P}^n)$ enters decisively. Consider the coordinate neighborhood $N \subset \mathbb{S}^{n-1}$ above and if $h \in C^\infty(\mathbb{R}^n - \{0\})$ let $h^*(\omega_1,\dots,\omega_{n-1},s)$ be the function obtained from h by means of the substitution (9). Then

$$\frac{\partial h}{\partial u_i} = \sum_{j=1}^{n-1} \frac{\partial h^*}{\partial \omega_j} \frac{\partial \omega_j}{\partial u_i} + \frac{\partial h^*}{\partial s} \cdot \frac{\partial s}{\partial u_i} \qquad (1 \leq i \leq n)$$

and

$$\frac{\partial \omega_j}{\partial u_j} = \frac{1}{s}(\delta_{ij} - \frac{u_i u_j}{s^2}) \qquad (1 \leq i \leq n, \qquad 1 \leq j \leq n-1) ,$$

$$\frac{\partial s}{\partial u_i} = \omega_i \qquad (1 \leq i \leq n-1), \quad \frac{\partial s}{\partial u_n} = (1 - \omega_1^2 - \ldots - \omega_{n-1}^2)^{\frac{1}{2}} .$$

Hence

$$\frac{\partial h}{\partial u_i} = \frac{1}{s}\frac{\partial h^*}{\partial \omega_i} + \omega_i \left(\frac{\partial h^*}{\partial s} - \frac{1}{s}\sum_{j=1}^{n-1} \omega_j \frac{\partial h^*}{\partial \omega_j}\right) \qquad (1 \leq i \leq n-1)$$

$$\frac{\partial h}{\partial u_n} = (1 - \omega_1^2 - \ldots - \omega_1^2)^{\frac{1}{2}} \left(\frac{\partial h^*}{\partial s} - \frac{1}{s}\sum_{j=1}^{n-1} \omega_j \frac{\partial h^*}{\partial \omega_j}\right) .$$

In order to use this for $h = F$ we write

$$F(s\omega) = \int_{-\infty}^{\infty} \phi(\omega,r)dr + \int_{-\infty}^{\infty} \phi(\omega,r)(e^{-irs}-1)dr .$$

By assumption the first integral is independent of ω. Thus using (7) we have for a constant $K > 0$

$$\left| \frac{1}{s}\frac{\partial}{\partial \omega_i}(F(s\omega)) \right| \leq K \int (1 + r^4)^{-1} s^{-1} \left| e^{-isr}-1\right| dr \leq K \int \frac{|r|}{1+r^4} dr$$

and a similar estimate is obvious for $\partial F(s\omega)/\partial s$. The formulas above therefore imply that all the derivatives $\partial F/\partial u_i$ are bounded in a punctured ball $0 < |u| < \varepsilon$ so F is certainly continuous at $u = 0$.

More generally, we prove by induction that

$$(11) \quad \frac{\partial^q h}{\partial u_{i_1} \ldots \partial u_{i_q}} = \sum_{1 \leq i+j \leq q,\ 1 \leq k_1, \ldots k_i \leq n-1} A_{j,k_1 \ldots k_i}(\omega,s) \frac{\partial^{i+j} h^*}{\partial \omega_{k_1} \ldots \partial \omega_{k_i} \partial s^j}$$

where the A have the form

$$(12) \qquad A_{j,k_1\ldots k_i}(\omega,s) = a_{j,k_1\ldots k_i}(\omega)s^{j-q} .$$

For $q = 1$ this is in fact proved above. Assuming (11) for q we calculate

$$\frac{\partial^{q+1}h}{\partial u_{i_1} \ldots \partial u_{i_{q+1}}}$$

using the above formulas for $\partial/\partial u_i$. If $A_{j,k_1\ldots k_i}(\omega,s)$ is differentiated with respect to $u_{i_{q+1}}$ we get a formula like (12) with q replaced by $q + 1$. If on the other hand the $(i + j)^{th}$ derivative of h^* in (11) is differentiated with respect to $u_{i_{q+1}}$ we get a combination of terms

$$s^{-1} \frac{\partial^{i+j+1}h^*}{\partial\omega_{k_1} \ldots \partial\omega_{k_{i+1}}\partial s^j} \quad , \quad \frac{\partial^{i+j+1}h^*}{\partial\omega_{k_1} \ldots \partial\omega_{k_i}\partial s^{j+1}}$$

and in both cases we get coefficients satisfying (12) with q replaced by $q + 1$. This proves (11)-(12) in general. Now

$$(13) \quad F(s\omega) = \int_{-\infty}^{\infty} \phi(\omega,r) \sum_{0}^{q-1} \frac{(-isr)^k}{k!} dr + \int_{-\infty}^{\infty} \phi(\omega,r)e_q(-irs)dr,$$

where

$$e_q(t) = \frac{t^q}{q!} + \frac{t^{q+1}}{(q+1)!} + \ldots$$

Our assumption on ϕ implies that the first integral in (13) is a polynomial in $u_1,\ldots,u_n$ of degree $\leq q-1$ and is therefore annihilated by the differential operator (11). If $0 \leqslant j \leqslant q$, we have

$$(14) \qquad \left| s^{j-q} \frac{\partial^j}{\partial s^j}(e_q(-irs)) \right| = \left| (-ir)^q(-irs)^{j-q} e_{q-j}(-irs) \right| \leq k_j r^q ,$$

where k_j is a constant because the function $t \to (it)^{-p} e_p(it)$ is obviously bounded on $\mathbb{R}$ $(p \geq 0)$. Since $\phi \in \mathcal{S}(\mathbb{P}^n)$ it follows from (11)-(14) that each qth order derivative of F with respect to $u_1,\ldots,u_n$ is bounded in a punctured ball $0 < |u| < \varepsilon$. Thus we have proved $F \in C^\infty(\mathbb{R}^n)$. That F is rapidly decreasing is now clear form (7) and (11). Finally, if f is the function in $\mathcal{S}(\mathbb{R}^n)$ whose Fourier transform is F then

$$\tilde{f}(s\omega) = F(s\omega) = \int_{-\infty}^{\infty} \phi(\omega,r)e^{-irs}dr;$$

hence by (4), $\hat{f} = \phi$ and the theorem is proved.

To make further progress we introduce some useful notation. Let $S^r(x)$ denote the sphere $\{y : |y-x| = r\}$ in $\mathbb{R}^n$ and $A(r)$ its area. Let $B^r(x)$ denote the open ball $\{y : |y-x| < r\}$. For a continuous function f on $S^r(x)$ let $(M^r f)(x)$ denote the mean value

$$(M^r f)(x) = \frac{1}{A(r)} \int_{S^r(x)} f(\omega)d\omega ,$$

where $d\omega$ is the Euclidean measure. Let K denote the orthogonal group $\mathbf{O}(n)$, dk its Haar measure, normalized by $\int dk = 1$. If $y \in \mathbb{R}^n$, $r = |y|$ then

$$(15) \qquad (M^r f)(x) = \int_K f(x + k\cdot y)dk \quad .$$

In fact, for x,y fixed both sides represent rotation-invariant functionals on $C(S^r(x))$, having the same value for the function $f \equiv 1$. The rotations being transitive on $S^r(x)$, (15) follows from the uniqueness of such invariant functionals. Formula (3) can similarly be written

$$\check{\phi}(x) = \int_K \phi(x + k\cdot\xi_0)dk \tag{16}$$

if ξ_0 is some fixed hyperplane through the origin. We see then that if $f \in \mathcal{S}(\mathbb{R}^n)$, Ω_k the area of the unit sphere in $\mathbb{R}^k$,

$$(\hat{f})^{\vee}(x) = \int_K \hat{f}(x + k\cdot\xi_0)dk = \int_K\left\{\int_{\xi_0} f(x + k\cdot y)dm(y)\right\}dk$$

$$= \int_{\xi_0} (M^{|y|}f)(x)\ dm(y) = \Omega_{n-1}\int_0^\infty r^{n-2}\left\{\frac{1}{\Omega_n}\int_{S^{n-1}} f(x + r\omega)d\omega\right\}dr$$

so

$$(\hat{f})^{\vee}(x) = \frac{\Omega_{n-1}}{\Omega_n}\int_{\mathbb{R}^n} |x - y|^{-1}f(y)dy. \tag{17}$$

We consider now the analog of Theorem 2.4 for the transform $\phi \to \check{\phi}$. But $\phi \in \mathcal{S}_H(\mathbb{P}^n)$ does not imply $\check{\phi} \in \mathcal{S}(\mathbb{R}^n)$. (If this were so and we by Theorem 2.4 write $\phi = \hat{f}$, $f \in \mathcal{S}(\mathbb{R}^n)$ then the inversion formula in Theorem 3.1 for $n = 3$ would imply $\int f(x)dx = 0$). On a smaller space we shall obtain a more satisfactory result.

Let $\mathcal{S}^*(\mathbb{R}^n)$ denote the space of all functions $f \in \mathcal{S}(\mathbb{R}^n)$ which are orthogonal to all polynomials, i.e.,

$$\int_{\mathbb{R}^n} f(x)\ P(x)dx = 0 \quad \text{for all polynomials } P.$$

Similarly, let $\mathcal{S}^*(\mathbb{P}^n) \subset \mathcal{S}(\mathbb{P}^n)$ be the space of ϕ satisfying

$$\int_{\mathbf{R}} \phi(\omega,r)p(r)dr = 0 \quad \text{for all polynomials} \quad p.$$

Note that under the Fourier transform the space $\mathcal{S}^*(\mathbf{R}^n)$ corresponds to the subspace $\mathcal{S}_o(\mathbf{R}^n) \subset \mathcal{S}(\mathbf{R}^n)$ of functions all of whose derivatives vanish at 0.

COROLLARY 2.5. <u>The transforms</u> $f \to \hat{f}$, $\phi \to \check{\phi}$ <u>are bijections of</u> $\mathcal{S}^*(\mathbf{R}^n)$ <u>onto</u> $\mathcal{S}^*(\mathbf{P}^n)$ <u>and of</u> $\mathcal{S}^*(\mathbf{P}^n)$ onto $\mathcal{S}^*(\mathbf{R}^n)$, <u>respectively.</u>

The first statement is clear from (8) if we take into account the elementary fact that the polynomials $x \to (x,\omega)^k$ span the space of homogeneous polynomials of degree k. To see that $\phi \to \check{\phi}$ is a bijection of $\mathcal{S}^*(\mathbf{P}^n)$ onto $\mathcal{S}^*(\mathbf{R}^n)$ we use (17), knowing that $\phi = \hat{f}$ for some $f \in \mathcal{S}^*(\mathbf{R}^n)$. The right hand side of (17) is the convolution of f with the tempered distribution $|\mathbf{x}|^{-1}$ whose Fourier transform is by Lemma 8.2 a constant multiple of $|u|^{1-n}$. (Here we leave out the trivial case $n = 1$). By the general theory of tempered distributions (Schwartz [1966], Ch. VII, §8) this convolution is a tempered distribution whose Fourier transform is a constant multiple of $|u|^{1-n}\tilde{f}(u)$. But this lies in the space $\mathcal{S}_o(\mathbf{R}^n)$ since $\tilde{f}$ does. Now (17) implies that $\check{\phi} = (\hat{f})^{\vee} \in \mathcal{S}^*(\mathbf{R}^n)$ and that $\check{\phi} \not\equiv 0$ if $\phi \not\equiv 0$. Finally we see that the mapping $\phi \to \check{\phi}$ is surjective because the function $((\hat{f})^{\vee})^{\sim}(u) = c|u|^{1-n}\tilde{f}(u)$ (where c is a constant) runs through $\mathcal{S}_o(\mathbf{R}^n)$ as f runs through $\mathcal{S}^*(\mathbf{R}^n)$.

We now turn to the space $\mathcal{D}(\mathbf{R}^n) = C_c^\infty(\mathbf{R}^n)$ and its image under the Radon transform. We begin with a preliminary result.

THEOREM 2.6. (*The support theorem*). *Let* $f \in C(\mathbb{R}^n)$ *satisfy the following conditions*:

(i) *For each integer* $k > 0$, $|x|^k f(x)$ *is bounded*.

(ii) *There exists a constant* $A > 0$ *such that*

$$\hat{f}(\xi) = 0 \quad \text{for} \quad d(0,\xi) > A,$$

d *denoting distance*.

Then

$$f(x) = 0 \quad \text{for} \quad |x| > A.$$

Proof. Replacing f by the convolution $\phi * f$ where ϕ is a radial C^∞ function with support in a small ball $B^\varepsilon(0)$ we see that it suffices to prove the theorem for $f \in C^\infty(\mathbb{R}^n)$. In fact, $\phi * f$ is smooth, it satisfies (i) and by (5) it satisfies (ii) with A replaced by $A + \varepsilon$. Assuming the theorem for the smooth case we deduce that support $(\phi * f) \subset B^{A+\varepsilon}(0)$ so letting $\varepsilon \to 0$ we obtain support $(f) \subset$ Closure $B^A(0)$.

To begin with we assume f is a radial function. Then $f(x) = F(|x|)$ where $F \in C^\infty(\mathbb{R})$ and even. Then $\hat{f}$ has the form $\hat{f}(\xi) = \hat{F}(d(0,\xi))$ where $\hat{F}$ is given by

$$\hat{F}(p) = \int_{\mathbb{R}^{n-1}} F((p^2 + |y|^2)^{\frac{1}{2}})\,dm(y), \qquad (p \geq 0)$$

because of the definition of the Radon transform. In particular, F being even, $\hat{F}$ extends to an even function in $C^\infty(\mathbb{R})$. Using polar coordinates in $\mathbb{R}^{n-1}$ we obtain

(18) $$\hat{F}(p) = \Omega_{n-1}\int_0^{\infty} F((p^2+t^2)^{\frac{1}{2}})t^{n-2}dt.$$

Here we substitute $s = (p^2+t^2)^{-\frac{1}{2}}$ and then put $u = p^{-1}$. Then (18) becomes

$$u^{n-3}\hat{F}(u^{-1}) = \Omega_{n-1}\int_0^u (F(s^{-1})s^{-n})(u^2-s^2)^{\frac{1}{2}(n-3)}ds.$$

We write this equation for simplicity

(19) $$h(u) = \int_0^u g(s)(u^2-s^2)^{\frac{1}{2}(n-3)}ds \ .$$

This integral equation is very close to Abel's integral equation (Whittaker-Watson [1927], Ch. IX) and can be inverted as follows. Multiplying both sides by $u(t^2-u^2)^{\frac{1}{2}(n-3)}$ and integrating over $0 \leqslant u \leqslant t$ we obtain

$$\int_0^t h(u)(t^2-u^2)^{\frac{1}{2}(n-3)}udu =$$

$$\int_0^t \left(\int_0^u g(s)\left[(u^2-s^2)(t^2-u^2)\right]^{\frac{1}{2}(n-3)}ds\right)udu$$

$$= \int_0^t g(s)\left(\int_{u=s}^t u\ \left[(t^2-u^2)(u^2-s^2)\right]^{\frac{1}{2}(n-3)}du\right)ds \ .$$

The substitution $(t^2-s^2)V = (t^2+s^2) - 2u^2$ gives an explicit evaluation of the inner integral and we obtain

$$\int_0^t h(u)(t^2-u^2)^{\frac{1}{2}(n-3)}udu = C\int_0^t g(s)(t^2-s^2)^{n-2}ds.$$

Here we apply the operator $\frac{d}{d(t^2)} = \frac{1}{2t}\frac{d}{dt}$ (n-1) times whereby the right hand side gives a constant multiple of $t^{-1}g(t)$. Hence we obtain

$$(20) \qquad F(t^{-1})t^{-n} = ct\left(\frac{d}{d(t^2)}\right)^{n-1}\int_0^t (t^2 - u^2)^{\frac{1}{2}(n-3)}u^{n-2}\hat{F}(u^{-1})du.$$

By assumption (ii) we have $\hat{F}(u^{-1}) = 0$ if $u^{-1} \geq A$, that is if $u \leq A^{-1}$. But then (20) implies $F(t^{-1}) = 0$ if $t \leq A^{-1}$, that is if $t^{-1} \geq A$. This proves the theorem for the case when f is radial.

We consider next the case of a general f. Fix $x \in \mathbb{R}^n$ and consider the function

$$g_x(y) = \int_K f(x + k\cdot y)dk$$

as in (15). Then g_x satisfies (i) and

$$(21) \qquad \hat{g}_x(\xi) = \int_K \hat{f}(x + k\cdot\xi)dk,$$

$x + k\cdot\xi$ denoting the translate of the hyperplane $k\cdot\xi$ by x. The triangle inequality shows that

$$d(0, x+ k\cdot\xi) \geq d(0,\xi)-|x|, \qquad x \in \mathbb{R}^n, \quad k \in K.$$

Hence we conclude from assumption (i) and (21) that

$$(22) \qquad \hat{g}_x(\xi) = 0 \qquad \text{if} \qquad d(0,\xi) > A + |x|.$$

But g_x is a radial function so (22) implies by the first part of the proof that

$$(23) \qquad \int_K f(x + k\cdot y)dk = 0 \qquad \text{if} \quad |y| > A + |x|.$$

Geometrically, this formula reads: The surface integral of f over $S^{|y|}(x)$ is 0 if the ball $B^{|y|}(x)$ contains the ball $B^A(0)$. The theorem is therefore a consequence of the following lemma.

LEMMA 2.7. Let $f \in C(\mathbb{R}^n)$ be such that for each integer $k > 0$,

$$\sup_{x \in \mathbb{R}^n} |x|^k |f(x)| < \infty.$$

Suppose f has surface integral 0 over every sphere S which encloses the unit ball. Then $f(x) \equiv 0$ for $|x| > 1$.

Proof. The idea is to perturb S in the relation

$$(24) \qquad \int_S f(s)d\omega(s) = 0$$

slightly, and differentiate with respect to the parameter of perturbations, thereby obtaining additional relations. Replacing, as above, f with a suitable convolution $\phi * f$ we see that it suffices to prove the lemma for $f \in C^\infty(\mathbb{R}^n)$. Writing $S = S^R(x)$ and viewing the exterior of the ball $B^R(x)$ as a union of spheres with center x we have by the assumptions,

$$\int_{B^R(x)} f(y)dy = \int_{\mathbb{R}^n} f(y)dy,$$

which is a constant. Differentiating with respect to x_i we obtain

$$(25) \qquad \int_{B^R(0)} (\partial_i f)(x + y)dy = 0.$$

We use now the divergence theorem

$$\int_{B^R(0)} (\mathrm{div}F)(y)dy = \int_{S^R(0)} (F,\bar{n})(s)d\omega(s) \tag{26}$$

for a vector field F on $\mathbb{R}^n$, $\bar{n}$ denoting the outgoing unit normal and $d\omega$ the surface element on $S^R(0)$. For the vector field $F(y) = f(x+y)\frac{\partial}{\partial y_i}$ we obtain from (25) and (26), since $\bar{n} = R^{-1}(s_1,\ldots,s_n)$,

$$\int_{S^R(0)} f(x+s)\, s_i\, d\omega(s) = 0. \tag{27}$$

But by (24)

$$\int_{S^R(0)} f(x+s)\, x_i\, d\omega(s) = 0$$

so by adding

$$\int_S f(s)\, s_i\, d\omega(s) = 0.$$

This means that the hypotheses of the lemma hold for $f(x)$ replaced by the function $x_i f(x)$. By iteration

$$\int_S f(s)P(s)d\omega(s) = 0$$

for any polynomial P, so $f \equiv 0$ on S. This proves the lemma as well as Theorem 2.6.

COROLLARY 2.8. *Let* $f \in C(\mathbb{R}^n)$ *satisfy* (i) *in Theorem* 2.6 *and assume*

$$\hat{f}(\xi) = 0$$

for all hyperplanes ξ *disjoint from a certain compact convex set* C. *Then*

(28) $$f(x) = 0 \quad \underline{\text{for}} \quad x \notin C.$$

In fact, if B is a closed ball containing C we have by Theorem 2.6, $f(x) = 0$ for $x \notin B$. But C is the intersection of such balls so (28) follows.

<u>Remark 2.9.</u> While condition (i) of rapid decrease entered in the proof of Lemma 2.7 (we used $|x|^k f(x) \in L^1(\mathbb{R}^n)$ for each $k > 0$) one may wonder whether it could not be weakened in Theorem 2.6 and perhaps even dropped in Lemma 2.7.

As an example, showing that the condition of rapid decrease can not be dropped in either result consider for $n = 2$ the function

$$f(x,y) = (x + iy)^{-5}$$

made smooth in $\mathbf{R}^2$ by changing it in a small disk around 0. Using Cauchy's theorem for a large semicircle we have $\int_\ell f(x)dm(x) = 0$ for every line ℓ outside the unit circle. Thus (ii) is satisfied in Theorem 2.6. Hence (i) cannot be dropped or weakened substantially.

This same example works for Lemma 2.7. In fact, let S be a circle $|z - z_0| = r$ enclosing the unit disk. Then $d\omega(s) = -ir\frac{dz}{z-z_0}$ and by residue calculus,

$$\int_S z^{-5}(z - z_0)^{-1}dz = 0,$$

(the residues at $z = 0$ and $z = z_0$ cancel) so we have in fact

$$\int_S f(s)\,d\omega(s) = 0.$$

We recall now that $\mathcal{D}_H(\mathbb{P}^n)$ is the space of C^∞ - functions $\phi(\xi) = \phi(\omega,p)$ on $\mathbb{P}^n$ of compact support such that for each $k \in \mathbb{Z}^+$ $\int_{\mathbb{R}} \phi(\omega,p)p^k dp$ is a homogeneous k^{th} degree polynomial in $\omega_1,\ldots,\omega_n$. Combining Theorems 2.4, 2.6 we obtain the following characterization of the Radon transform of the space $\mathcal{D}(\mathbb{R}^n) = C_c^\infty(\mathbb{R}^n)$. This can be regarded as the analog for the Radon transform of the Paley-Wiener theorem for the Fourier transform (cf. Hörmander [1963]).

THEOREM 2.10. (*The Paley-Wiener theorem*). *The Radon transform is a bijection of* $\mathcal{D}(\mathbb{R}^n)$ *onto* $\mathcal{D}_H(\mathbb{P}^n)$.

We conclude this section with a variation and a consequence of Theorem 2.6.

LEMMA 2.11. *Let* $f \in C_c(\mathbb{R}^n)$, $A > 0$, ω_0 *a fixed unit vector and* $N \subset S$ *a neighborhood of* ω_0 *in the unit sphere* $S \subset \mathbb{R}^n$. *Assume*

$$\hat{f}(\omega,p) = 0 \qquad \textit{for} \qquad \omega \in N,\ \ p > A.$$

Then

(29) $$f(x) = 0 \ \ \textit{in the half-space} \ \ (x,\omega_0) > A.$$

Proof. Let B be a closed ball around the origin containing the support of f. Let $\varepsilon > 0$ and let H_ε be the union of the half spaces $(x,\omega) > A + \varepsilon$ as ω runs through N. Then by our assumption

$$(30) \qquad \hat{f}(\xi) = 0 \qquad \text{if} \qquad \xi \subset H_\varepsilon .$$

Now choose a ball B_ε with a center on the ray from 0 through $-\omega_0$, with the point $(A + 2\varepsilon)\omega_0$ on the boundary, and with radius so large that any hyperplane ξ intersecting B but not B_ε must be in H_ε. Then by (30)

$$\hat{f}(\xi) = 0 \qquad \text{whenever} \qquad \xi \in \mathbf{P}^n,\ \xi \cap B_\varepsilon = \emptyset.$$

Hence by Theorem 2.6, $f(x) = 0$ for $x \notin B_\varepsilon$. In particular, $f(x) = 0$ for $(x,\omega_0) > A + 2\varepsilon$; since $\varepsilon > 0$ is arbitrary, the lemma follows.

COROLLARY 2.12. _Let_ N _be any open subset of the unit sphere_ $\mathbf{S}^{n-1}$. _If_ $f \in C_c(\mathbb{R}^n)$ _and_

$$\hat{f}(\omega,p) = 0 \qquad \underline{\text{for}} \qquad p \in \mathbb{R},\ \omega \in N$$

then $f \equiv 0$.

Since $\hat{f}(-\omega,-p) = \hat{f}(\omega,p)$ this is obvious from Lemma 2.11.

§3. The Inversion Formulas.

We shall now establish explicit inversion formulas for the Radon transform $f \to \hat{f}$ and its dual $\phi \to \check{\phi}$.

THEOREM 3.1. _The function_ f _can be recovered from its Radon transform by means of the following inversion formula:_

$$cf = L^{\frac{1}{2}(n-1)}\left((\hat{f})^\vee\right) \qquad f \in \mathcal{S}(\mathbb{R}^n)$$

where c *is the constant*

$$c = (-4\pi)^{\frac{n-1}{2}} \frac{\Gamma(\frac{n}{2})}{\Gamma(\frac{1}{2})} .$$

Here L is the Laplacian on $\mathbb{R}^n$. For n even the fractional power $L^{\frac{n-1}{2}}$ requires a definition which will be given in course of the proof.

To begin with we recall the familiar fact that if $f \in C^2(\mathbb{R}^n)$ is a radial function, i.e., $f(x) = F(r)$, $r = |x|$, then

(31) $$(Lf)(x) = \frac{d^2F}{dr^2} + \frac{n-1}{r}\frac{dF}{dr} .$$

This is immediate from the relations

$$\frac{\partial^2 f}{\partial x_i^2} = \frac{\partial^2 f}{\partial r^2}\left(\frac{\partial r}{\partial x_i}\right)^2 + \frac{\partial f}{\partial r}\frac{\partial^2 r}{\partial x_i^2} .$$

LEMMA 3.2.

(i) $LM^r = M^rL$ *for each* $r > 0$.

(ii) *For* $f \in C^2(\mathbb{R}^n)$ *the mean value* $(M^rf)(x)$ *satisfies the "Darboux equation"*

$$L_x((M^rf)(x)) = \left(\frac{\partial^2}{\partial r^2} + \frac{n-1}{r}\frac{\partial}{\partial r}\right)(M^rf(x)),$$

that is, the function $F(x,y) = (M^{|y|}f)(x)$ *satisfies*

$$L_x(F(x,y)) = L_y(F(x,y)).$$

Proof. We prove this group theoretically, using expression (15) for the mean value. For $z \in \mathbf{R}^n$, $k \in K$ let T_z denote the translation $x \to x + z$ and R_k the rotation $x \to k\cdot x$. Since L is invariant under these transformations, we have if $r = |y|$,

$$(LM^r f)(x) = \int_K L_x(f(x + k\cdot y)dk = \int_K (Lf)(x + k\cdot y)dk$$

$$= (M^r Lf)(x) = \int_K \left[(Lf) \circ T_x \circ R_k \right](y)dk$$

$$= \int_K \left[L(f \circ T_x \circ R_k) \right](y)dk = L_y(\int_K f(x + k\cdot y)dk)$$

which proves the lemma.

In order to prove Theorem 3.1 let $f \in \mathcal{S}(\mathbf{R}^n)$. Fix a hyperplane ξ_0 through 0, and an isometry $g \in M(n)$. As k runs through $O(n)$, $gk\cdot\xi_0$ runs through the set of hyperplanes through $g\cdot 0$, and we have

$$\check{\phi}(g\cdot 0) = \int_K \phi(gk\cdot\xi_0)dk$$

so

$$(\hat{f})^{\vee}(g\cdot 0) = \int_K \left(\int_{\xi_0} f(gk\cdot y)dm(y) \right) dk$$

$$= \int_{\xi_0} dm(y) \int_K f(gk\cdot y)dk = \int_{\xi_0} (M^{|y|} f)(g\cdot 0)dm(y) .$$

Hence

$$(32)\qquad \left((\hat{f})\right)^{\vee}(x) = \Omega_{n-1}\int_0^{\infty} (M^r f)(x) r^{n-2} dr,$$

where Ω_{n-1} is the area of the unit sphere in $\mathbb{R}^{n-1}$, i.e.

$$\Omega_{n-1} = \frac{2\pi^{\frac{n-1}{2}}}{\Gamma(\frac{n-1}{2})}.$$

Applying L to (32), using (31) and Lemma 3.2, we obtain

$$(33)\qquad L\left((\hat{f})^{\vee}\right) = \Omega_{n-1}\int_0^{\infty} \left(\frac{d^2F}{dr^2} + \frac{n-1}{r}\frac{dF}{dr}\right) r^{n-2}\, dr$$

where $F(r) = (M^r f)(x)$. Integrating by parts and using $F(o) = f(x)$, $\lim_{r\to\infty} r^k F(r) = 0$, we get

$$L\left[(\hat{f})^{\vee}\right] = \begin{cases} -\Omega_{n-1} f(x) & \text{if} \quad n = 3, \\ -\Omega_{n-1}(n-3)\int_0^{\infty} F(r) r^{n-4} dr & (n > 3). \end{cases}$$

More generally,

$$L_x\left[\int_0^{\infty} (M^r f)(x) r^k dr\right] = \begin{cases} -(n-2) f(x) & \text{if} \quad k = 1, \\ -(n-1-k)(k-1)\int_0^{\infty} F(r) r^{k-2} dr, & (k > 1). \end{cases}$$

If n is odd the formula in Theorem 3.1 follows by iteration.

We now pass to the case of even n and use the definition of the fractional power $L^{\frac{n-1}{2}}$ in terms of Riesz potentials I^{γ} in (85), (86) §8. Using (17) in the present section we have

$$(34)\qquad (\hat{f})^{\vee} = 2^{n-1}\pi^{\frac{n}{2}-1}\,\Gamma(\tfrac{n}{2}) I^{n-1} f.$$

Using Prop. 8.6 and the definition of $L^{\frac{n-1}{2}}$ ((85) §8) we do obtain the desired formula

$$L^{\frac{n-1}{2}}\left((\hat{f})^{\vee}\right) = c\,f \tag{35}$$

in Theorem 3.1.

We shall now prove a similar inversion formula for the dual transform $\phi \to \check{\phi}$ on the subspace $\mathcal{S}^*(\mathbb{P}^n)$.

THEOREM 3.3. *We have*

$$c\phi = \Box^{\frac{n-1}{2}}\left((\check{\phi})^{\wedge}\right), \qquad \phi \in \mathcal{S}^*(\mathbb{P}^n),$$

where c *is the constant*

$$c = (-4\pi)^{\frac{n-1}{2}} \frac{\Gamma(\frac{n}{2})}{\Gamma(\frac{1}{2})}.$$

Here $\Box$ denotes as before the operator $\frac{d^2}{dp^2}$ and its fractional powers are again defined in terms of the Riesz' potentials on the one-dimensional p-space.

If n is odd our inversion formula follows from the odd-dimensional case in Theorem 3.1 if we put $f = \check{\phi}$ and take Lemma 2.1 and Corollary 2.5 into account. Suppose now n is even. We claim that

$$\left((-L)^{\frac{n-1}{2}} f\right)^{\wedge} = (-\Box)^{\frac{n-1}{2}} \hat{f} \qquad f \in \mathcal{S}^*(\mathbb{R}^n). \tag{36}$$

In fact, by Lemma 8.5 and Cor. 2.5 both sides belong to $\mathcal{S}^*(\mathbb{P}^n)$.

Taking the 1-dimensional Fourier transform of $\left[(-L)^{\frac{n-1}{2}} f\right]^{\wedge}$ we get by (4), $\left[(-L)^{\frac{n-1}{2}} f\right]^{\sim}(s\omega) = |s|^{n-1}\tilde{f}(s\omega)$. This coincides with the Fourier transform of $(-\Box)^{\frac{n-1}{2}} \hat{f}$ so (36) is proved. Now Theorem 3.3 follows from (35) if we put in (36)

$$\phi = \hat{g}, \qquad f = (\hat{g})^{\vee}, \qquad g \in \mathcal{S}^*(\mathbb{R}^n).$$

Because of its theoretical importance we now prove the inversion theorem (3.1) in a different form. The proof is less geometric and involves just the one variable Fourier transform.

Let $\mathcal{H}$ denote the Hilbert transform

$$(\mathcal{H}F)(t) = \frac{i}{\pi}\int_{-\infty}^{\infty} \frac{F(p)}{t-p}dp \qquad F \in \mathcal{S}(\mathbb{R})$$

the integral being considered as the Cauchy principal value. For $\phi \in \mathcal{S}(\mathbb{P}^n)$ let $\Lambda\phi$ be defined by

$$(37) \qquad (\Lambda\phi)(\omega,p) = \begin{cases} \dfrac{d^{n-1}}{dp^{n-1}}\phi(\omega,p) & n \text{ odd}, \\ \mathcal{H}_p \dfrac{d^{n-1}}{dp^{n-1}}\phi(\omega,p) & n \text{ even}. \end{cases}$$

Note that in both cases $(\Lambda\phi)(-\omega,-p) = (\Lambda\phi)(\omega,p)$ so $\Lambda\phi$ is a function on $\mathbb{P}^n$.

THEOREM 3.4. Let Λ be as defined by (37). Then

$$cf = (\Lambda\hat{f})^{\vee}, \qquad f \in \mathcal{S}(\mathbb{R}^n),$$

where as before

$$c = (-4\pi)^{\frac{n-1}{2}} \frac{\Gamma(\frac{n}{2})}{\Gamma(\frac{1}{2})} .$$

Proof. By the inversion formula for the Fourier transform and by (4)

$$f(x) = (2\pi)^{-n} \int_{\mathbf{S}^{n-1}} d\omega \int_0^\infty \left(\int_{-\infty}^\infty e^{-isp} \hat{f}(\omega,p)\,dp \right) e^{is(x,\omega)} s^{n-1} ds$$

which we write as

$$f(x) = (2\pi)^{-n} \int_{\mathbf{S}^{n-1}} F(\omega,x)\,d\omega = (2\pi)^{-n} \int_{\mathbf{S}^{n-1}} \tfrac{1}{2}\left(F(\omega,x) + F(-\omega,x)\right) d\omega$$

Using $\hat{f}(-\omega,-p) = \hat{f}(\omega,p)$ this gives the formula

$$\text{(38)} \qquad f(x) = \tfrac{1}{2}(2\pi)^{-n} \int_{\mathbf{S}^{n-1}} d\omega \int_{-\infty}^\infty |s|^{n-1} e^{is(x,\omega)} ds \int_{-\infty}^\infty e^{-isp}\, \hat{f}(\omega,p)\,dp \ .$$

If n is odd the absolute value on s can be dropped. The factor s^{n-1} can be removed by replacing $\hat{f}(\omega,p)$ by $(-i)^{n-1} \frac{d^{n-1}}{dp^{n-1}} \hat{f}(\omega,p)$. The inversion formula for the Fourier transform on $\mathbb{R}$ then gives

$$f(x) = \tfrac{1}{2}(2\pi)^{-n}(2\pi)^{+1}(-i)^{n-1} \int_{\mathbf{S}^{n-1}} \left\{ \frac{d^{n-1}}{dp^{n-1}} \hat{f}(\omega,p) \right\}_{p=(x,\omega)} d\omega$$

as desired.

Supposing now n is even we let

$$\operatorname{sgn} s = \begin{cases} 1 & \text{if} \quad s \geq 0 \\ -1 & \text{if} \quad s < 0 . \end{cases}$$

Then for a constant c_o,

(39) $$f(x) = c_o \int_{\mathbf{S}^{n-1}} d\omega \int (\operatorname{sgn} s)\, e^{is(x,\omega)} ds \int \frac{d^{n-1}}{dp^{n-1}} \hat{f}(\omega,p)\, e^{-isp}\, dp.$$

The Cauchy principal value

$$\psi \longrightarrow \lim_{\varepsilon \to 0} \int_{x \geq \varepsilon} \frac{\psi(x)}{x} dx$$

is a tempered distribution whose Fourier transform is $-\pi i \operatorname{sgn} s$ (cf. Schwartz [1966], Ch. VII). Hence

(40) $$(\mathcal{H}F)^{\sim}(s) = \operatorname{sgn} s\, \tilde{F}(s).$$

Thus $\operatorname{sgn} s$ can be removed in the integral above if we replace $\frac{d^{n-1}}{dp^{n-1}} \hat{f}(\omega,p)$ by $\Lambda\hat{f}(\omega,p)$. The inversion formula for the one-dimensional Fourier transform now gives the result.

§4. The Plancherel Formula.

We recall that the functions on $\mathbf{P}^n$ have been identified with the functions ϕ on $\mathbf{S}^{n-1} \times \mathbf{R}$ which are even: $\phi(-\omega,-p) = \phi(\omega,p)$. The functional

(41) $$\phi \longrightarrow \int_{\mathbf{S}^{n-1}} \int_{\mathbf{R}} \phi(\omega,p)\, d\omega dp, \quad \phi \in C_c(\mathbf{P}^n),$$

is therefore a well defined measure on $\mathbf{P}^n$, denoted $d\omega\, dp$. The group $M(n)$ of rigid motions of $\mathbf{R}^n$ acts transitively on $\mathbf{P}^n$; it also leaves the measure $d\omega dp$ invariant. It suffices to verify this latter statement for the translations T in $M(n)$ because $M(n)$ is generated by them together with the rotations around 0, and these rotations clearly leave $d\omega dp$ invariant. But

$$(\phi \circ T)(\omega,p) = \phi(\omega,p + q(\omega,T))$$

where $q(\omega,T) \in \mathbb{R}$ is independent of p so

$$\iint(\phi \circ T)(\omega,p)d\omega dp = \iint\phi(\omega,p + q(\omega,T))dpd\omega = \iint\phi(\omega,p)dpd\omega,$$

proving the invariance.

In accordance with (85),(86) in §8 the fractional power $\square^k$ is defined on $\mathcal{S}(\mathbb{P}^n)$ by

$$(42) \qquad (-\square)^k\phi(\omega,p) = \frac{1}{H_1(-2k)}\int_{\mathbb{R}}\phi(\omega,q)|p-q|^{-2k-1}dq$$

and then the one-dimensional Fourier transform satisfies

$$(43) \qquad \left((-\square)^k\phi\right)^{\sim}(\omega,s) = |s|^{2k}\tilde{\phi}(\omega,s).$$

Now, if $f \in \mathcal{S}(\mathbb{R}^n)$ we have by (4)

$$\hat{f}(\omega,p) = (2\pi)^{-1}\int\tilde{f}(s\omega)e^{isp}ds$$

and

$$(44) \qquad (-\square)^{\frac{n-1}{4}}\hat{f}(\omega,p) = (2\pi)^{-1}\int_{\mathbb{R}}|s|^{\frac{n-1}{2}}\tilde{f}(s\omega)e^{isp}ds.$$

THEOREM 4.1. *The mapping* $f \to \Box^{\frac{n-1}{4}} \hat{f}$ *extends to an isometry of* $L^2(\mathbb{R}^n)$ *onto the space* $L^2_e(\mathbb{S}^{n-1} \times \mathbb{R})$ *of even functions in* $L^2(\mathbb{S}^{n-1} \times \mathbb{R})$, *the measure on* $\mathbb{S}^{n-1} \times \mathbb{R}$ *being*

$$\frac{1}{2}(2\pi)^{1-n} d\omega dp.$$

Proof. By (44) we have from the Plancherel formula on $\mathbb{R}$

$$(2\pi)\int_{\mathbb{R}} \left|(-\Box)^{\frac{n-1}{4}} \hat{f}(\omega,p)\right|^2 dp = \int_{\mathbb{R}} \left|\,|s|^{n-1}\right| \tilde{f}(s\omega)|^2 ds$$

so by integration over $\mathbb{S}^{n-1}$ and using the Plancherel formula for $f(x) \to \tilde{f}(s\omega)$ we obtain

$$\int_{\mathbb{R}^n} |f(x)|^2\, dx = \frac{1}{2}(2\pi)^{1-n} \int_{\mathbb{S}^{n-1} \times \mathbb{R}} \left| \Box^{\frac{n-1}{4}} \hat{f}(\omega,p) \right|^2 d\omega dp.$$

It remains to prove that the mapping is surjective. For this it would suffice to prove that if $\phi \in L^2(\mathbb{S}^{n-1} \times \mathbb{R})$ is even and satisfies

$$\int_{\mathbb{S}^{n-1}} \int_{\mathbb{R}} \phi(\omega,p)(-\Box^{\frac{n-1}{4}} \hat{f}(\omega,p) d\omega dp = 0$$

for all $f \in \mathcal{S}(\mathbb{R}^n)$ then $\phi = 0$. Taking Fourier transforms we must prove that if $\psi \in L^2(\mathbb{S}^{n-1} \times \mathbb{R})$ is even and satisfies

$$\int_{\mathbb{S}^{n-1}} \int_{\mathbb{R}} \psi(\omega,s)|s|^{\frac{n-1}{2}} \tilde{f}(s\omega) ds d\omega = 0 \qquad (45)$$

for all $f \in \mathcal{S}(\mathbb{R}^n)$ then $\psi = 0$. Using the condition $\psi(-\omega,-s) = \psi(\omega,s)$ we see that

$$\int_{\mathbf{S}^{n-1}}\int_{-\infty}^{0}\psi(\omega,s)|s|^{\frac{1}{2}(n-1)}\tilde{f}(s\omega)\,ds\,d\omega$$
$$=\int_{\mathbf{S}^{n-1}}\int_{0}^{\infty}\psi(\omega,t)|t|^{\frac{1}{2}(n-1)}\tilde{f}(t\omega)\,dt\,d\omega$$

so (45) holds with $\mathbf{R}$ replaced with the positive axis $\mathbf{R}^+$. But then the function

$$\Psi(u) = \psi\left(\frac{u}{|u|}, |u|\right)|u|^{-\frac{1}{2}(n-1)}, \qquad u\in\mathbf{R}^n - \{0\}$$

satisfies

$$\int_{\mathbf{R}^n}\Psi(u)\tilde{f}(u)\,du = 0, \qquad f\in\mathcal{S}(\mathbf{R}^n)$$

so $\Psi = 0$ almost everywhere, whence $\psi = 0$.

§5. Radon Transform of Distributions.

It will be proved in a general context in Ch. III (Prop. 2.2) that

$$\text{(46)}\qquad \int_{\mathbf{P}^n}\hat{f}(\xi)\phi(\xi)\,d\xi = \int_{\mathbf{R}^n} f(x)\check{\phi}(x)\,dx$$

for $f\in C_c(\mathbf{R}^n)$, $\phi\in C(\mathbf{P}^n)$ if $d\xi$ is a suitable fixed $M(n)$-invariant measure on $\mathbf{P}^n$. Thus $d\xi = \gamma\,d\omega\,dp$ where γ is a constant, independent of f and ϕ. With applications to distributions in mind we shall prove (46) in a somewhat stronger form.

LEMMA 5.1. Formula (46) holds (with $\hat{f}$ and $\check{\phi}$ existing almost everywhere) in the following two situations:

a) $f \in L^1(\mathbb{R}^n)$ vanishing outside a compact set; $\phi \in C(\mathbb{P}^n)$.

b) $f \in C_c(\mathbb{R}^n)$, ϕ locally integrable.

Also $d\xi = \Omega_n^{-1}\, d\omega dp$.

Proof. We shall use the Fubini theorem repeatedly both on the product $\mathbb{R}^n \times \mathbb{S}^{n-1}$ and on the product $\mathbb{R}^n = \mathbb{R} \times \mathbb{R}^{n-1}$. Since $f \in L^1(\mathbb{R}^n)$ we have for each $\omega \in \mathbb{S}^{n-1}$ that $\hat{f}(\omega,p)$ exists for almost all p and

$$\int_{\mathbb{R}^n} f(x)dx = \int_{\mathbb{R}} \hat{f}(\omega,p)dp.$$

We also conclude that $\hat{f}(\omega,p)$ exists for almost all $(\omega,p) \in \mathbb{S}^{n-1} \times \mathbb{R}$. Next we consider the measurable function $(x,\omega) \longrightarrow f(x)\phi(\omega,x))$ on $\mathbb{R}^n \times \mathbb{S}^{n-1}$. We have

$$\int_{\mathbb{S}^{n-1} \times \mathbb{R}^n} |f(x)\phi(\omega,(\omega,x))|\, d\omega dx$$

$$= \int_{\mathbb{S}^{n-1}} \left(\int_{\mathbb{R}^n} |f(x)\phi(\omega,(\omega,x))|\, dx \right) d\omega$$

$$= \int_{\mathbb{S}^{n-1}} \left(\int_{\mathbb{R}} |f|\hat{}(\omega,p)\, |\phi(\omega,p)|\, dp \right) d\omega,$$

which in both cases is finite. Thus $f(x)\cdot\phi(\omega,(\omega,x))$ is integrable on $\mathbb{R}^n \times \mathbb{S}^{n-1}$ and its integral can be calculated by removing the absolute values above. This gives the left hand side of (46). Reversing the integrations we conclude that $\check{\phi}(x)$ exists for almost all x and that the double integral reduces to the right hand side of (46).

The formula (46) dictates how to define the Radon transform and its dual for distributions. In order to make the definitions formally consistent with those for functions we would require $\hat{S}(\phi) = S(\check{\phi})$, $\check{\Sigma}(f) = \Sigma(\hat{f})$ if S and Σ are distributions on $\mathbb{R}^n$ and $\mathbb{P}^n$, respectively. But while $f \in \mathcal{D}(\mathbb{R}^n)$ implies $\hat{f} \in \mathcal{D}(\mathbb{P}^n)$ a similar implication does not hold for ϕ ; we do not even have $\check{\phi} \in \mathcal{S}(\mathbb{R}^n)$ for $\phi \in \mathcal{D}(\mathbb{P}^n)$ so $\hat{S}$ cannot be defined as above even if S is assumed to be tempered. Using the notation $\mathcal{E}$ (resp. $\mathcal{D}$) for the space of C^∞ functions (resp. of compact support) and $\mathcal{D}'$ (resp. $\mathcal{E}'$) for the space of distributions (resp. of compact support) we make the following definition.

<u>Definition</u>. For $S \in \mathcal{E}'(\mathbb{R}^n)$ we define the functional $\hat{S}$ by

$$\hat{S}(\phi) = S(\check{\phi}) \qquad \text{for} \qquad \phi \in \mathcal{E}(\mathbb{P}^n);$$

for $\Sigma \in \mathcal{D}'(\mathbb{P}^n)$ we define the functional $\check{\Sigma}$ by

$$\check{\Sigma}(f) = \Sigma(\hat{f}) \qquad \text{for} \qquad f \in \mathcal{D}(\mathbb{R}^n).$$

LEMMA 5.2.

(i) <u>For each</u> $\Sigma \in \mathcal{D}'(\mathbb{P}^n)$ <u>we have</u> $\check{\Sigma} \in \mathcal{D}'(\mathbb{R}^n)$.

(ii) <u>For each</u> $S \in \mathcal{E}'(\mathbb{R}^n)$ <u>we have</u> $\hat{S} \in \mathcal{E}'(\mathbb{P}^n)$.

<u>Proof</u>. For $A > 0$ let $\mathcal{D}_A(\mathbb{R}^n)$ denote the set of functions $f \in \mathcal{D}(\mathbb{R}^n)$ with support in the closure of $B^A(0)$. Similarly let $\mathcal{D}_A(\mathbb{P}^n)$ denote the set of functions $\phi \in \mathcal{D}(\mathbb{P}^n)$ with support in the closure of the "ball"

$$\beta^A(0) = \{\xi \in \mathbb{P}^n : d(0,\xi) < A\}.$$

The mapping $f \to \hat{f}$ from $\mathcal{D}_A(\mathbb{R}^n)$ to $\mathcal{D}_A(\mathbb{P}^n)$ being continuous (with the topologies defined in §8) the restriction of $\check{\Sigma}$ to each $\mathcal{D}_A(\mathbb{R}^n)$ is continuous so (i) follows. That $\hat{S}$ is a distribution is clear from (3). Concerning its support select $R > 0$ such that S has support inside $B^R(0)$. Then if $\phi(\omega,p) = 0$ for $|p| \leq R$ we have $\check{\phi}(x) = 0$ for $|x| \leq R$ whence $\hat{S}(\phi) = S(\check{\phi}) = 0$.

LEMMA 5.3. *For* $S \in \mathcal{E}'(\mathbb{R}^n)$, $\Sigma \in \mathcal{D}'(\mathbb{P}^n)$ *we have*

$$(LS)^{\wedge} = \Box\hat{S}, \qquad (\Box\Sigma)^{\vee} = L\check{\Sigma}$$

Proof. In fact by Lemma 2.1,

$$(LS)^{\wedge}(\phi) = (LS)(\check{\phi}) = S(L\check{\phi}) = S((\Box\phi)^{\vee}) = \hat{S}(\Box\phi) = (\Box\hat{S})(\phi).$$

The other relation is proved in the same manner.

We shall now prove an analog of the support theorem (Theorem 2.6) for distributions. For $A > 0$ let $\beta^A(0)$ be defined as above.

THEOREM 5.4. *Let* $T \in \mathcal{E}'(\mathbb{R}^n)$ *satisfy the condition*

$$\operatorname{supp}\hat{T} \subset C\ell(\beta^A(0)), \qquad (C\ell = \text{closure}).$$

Then

$$\operatorname{supp}(T) \subset C\ell(B^A(0)).$$

Proof. For $f \in \mathcal{D}(\mathbb{R}^n)$, $\phi \in \mathcal{D}(\mathbb{P}^n)$ we can consider the "convolution"

$$(f \times \phi)(\xi) = \int_{\mathbb{R}^n} f(y)\phi(\xi-y)dy,$$

where for $\xi \in \mathbb{P}^n$, $\xi-y$ denotes the translation of the hyperplane ξ by $-y$. Then

$$(f \times \phi)^{\vee} = f * \overset{\vee}{\phi}.$$

In fact, if ξ_0 is any hyperplane through 0,

$$(f \times \phi)^{\vee}(x) = \int_K dk \int_{\mathbb{R}^n} f(y)\phi(x + k\cdot\xi_0 - y)dy$$

$$= \int_K dk \int_{\mathbb{R}^n} f(x - y)\phi(y + k\cdot\xi_0)dy = (f * \overset{\vee}{\phi})(x).$$

By the definition of $\hat{T}$, the support assumption on $\hat{T}$ is equivalent to

$$T(\overset{\vee}{\phi}) = 0$$

for all $\phi \in \mathcal{D}(\mathbb{P}^n)$ with support in $\mathbb{P}^n - C\ell(\beta^A(0))$. Let $\varepsilon > 0$, let $f \in \mathcal{D}(\mathbb{R}^n)$ be a symmetric function with support in $C\ell(B^{\varepsilon}(0))$ and let $\phi \in \mathcal{D}(\mathbb{P}^n)$ have support contained in $\mathbb{P}^n - C\ell(\beta^{A+\varepsilon}(0))$. Since $d(0,\xi - y) \leq d(0,\xi) + |y|$ it follows that $f \times \phi$ has support in $\mathbb{P}^n - C\ell(\beta^A(0))$; thus by the formulas above, and the symmetry of f,

$$(f * T)(\overset{\vee}{\phi}) = T(f * \overset{\vee}{\phi}) = T((f \times \phi)^{\vee}) = 0.$$

But then

$$(f * T)^{\wedge}(\phi) = (f * T)(\check{\phi}) = 0,$$

which means that $(f * T)^{\wedge}$ has support in $C\ell(B^{A+\varepsilon}(0))$. But now Theorem 2.6 implies that $f * T$ has support in $C\ell(B^{A+\varepsilon}(0))$. Letting $\varepsilon \to 0$ we obtain the desired conclusion supp $(T) \subset C\ell(B^{A}(0))$.

We can now extend the inversion formulas for the Radon transform to distributions. First we observe that the Hilbert transform $\mathcal{H}$ can be extended to distributions T on $\mathbb{R}$ of compact support. It suffices to put

$$\mathcal{H}(T)(F) = T(\mathcal{H}F) \qquad F \in \mathcal{D}(\mathbb{R}).$$

In fact, $\mathcal{H}$ being the convolution with a tempered distribution the mapping $F \longrightarrow \mathcal{H}F$ is a continuous mapping of $\mathcal{D}(\mathbb{R})$ into $\mathcal{E}(\mathbb{R})$ (cf. Schwartz [1966], Ch. VII §5). In particular $\mathcal{H}(T) \in \mathcal{D}'(\mathbb{R})$.

THEOREM 5.5. *The Radon transform* $S \longrightarrow \hat{S}$ $(S \in \mathcal{E}'(\mathbb{R}^n))$ *is inverted by the following formula*

$$cS = (\Lambda\hat{S})^{\vee}, \qquad S \in \mathcal{E}'(\mathbb{R}^n),$$

where the constant c *equals*

$$c = (-4\pi)^{\frac{n-1}{2}} \frac{\Gamma(\frac{n}{2})}{\Gamma(\frac{1}{2})}.$$

In the case when n *is odd we have also*

$$cS = L^{\frac{1}{2}(n-1)}\left\{(\hat{S})^{\vee}\right\}.$$

Remark. Since $\hat{S}$ has compact support and since Λ is defined by means of the Hilbert transform the remarks above show that $\Lambda\hat{S} \in \mathcal{D}'(\mathbb{P}^n)$ so the right hand side is well defined.

Proof. Using Theorem 3.4 we have

$$(\Lambda\hat{S})^{\vee}(f) = (\Lambda\hat{S})(\hat{f}) = \hat{S}(\Lambda\hat{f}) = S((\Lambda\hat{f})^{\vee}) = cS(f).$$

The other inversion formula then follows, using the lemma.

Let M be a manifold and $d\mu$ a measure such that on each local coordinate patch with coordinates $(t_1,\dots,t_n)$ the Lebesque measure $dt_1,,,dt_n$ and $d\mu$ are absolutely continuous with respect to each other. If h is a function on M locally integrable with respect to $d\mu$ the distribution $\phi \to \int \phi h\,d\mu$ will be denoted by T_h.

PROPOSITION 5.6.

a) *Let $f \in L^1(\mathbb{R}^n)$ vanish outside a compact set. Then the distribution T_f has Radon transform given by*

(47)
$$\hat{T}_f = T_{\hat{f}}.$$

b) *Let ϕ be a locally integrable function on $\mathbb{P}^n$. Then*

(48)
$$(T_\phi)^{\vee} = T_{\check{\phi}}.$$

Proof. The existence and local integrability of $\hat{f}$ and $\check{\phi}$ was established during the proof of Lemma 5.1. The two formulas now follow directly from Lemma 5.1.

As a result of this proposition the smoothness assumption can be dropped in the inversion formula. In particular, we can state the following result.

COROLLARY 5.7. (n odd). *The inversion formula*

$$cf = L^{\frac{n-1}{2}}\left[(\hat{f})^{\vee}\right], \qquad c = (-4\pi)^{\frac{n-1}{2}} \frac{\Gamma(\frac{n}{2})}{\Gamma(\frac{1}{2})},$$

holds for all $f \in L^1(\mathbb{R}^n)$ *of compact support, the derivative interpreted in the sense of distributions.*

Examples. If μ is a measure (or a distribution) on a submanifold S of a manifold M the distribution on M given by $\phi \to \mu(\phi|S)$ will also be denoted by μ.

a) Let δ_0 be the delta distribution $f \to f(0)$ on $\mathbb{R}^n$. Then

$$\hat{\delta}_0(\phi) = \delta_0(\check{\phi}) = \Omega_n^{-1} \int_{\mathbf{S}^{n-1}} \phi(\omega, 0)\, d\omega$$

so

$$\hat{\delta}_o = \Omega_n^{-1} m_{\mathbf{S}^{n-1}} \tag{49}$$

the normalized measure on $\mathbf{S}^{n-1}$ considered as a distribution on $\mathbf{S}^{n-1} \times \mathbb{R}$.

b) Let ξ_0 denote the hyperplane $x_n = 0$ in $\mathbb{R}^n$, and δ_{ξ_0} the delta distribution $\phi \to \phi(\xi_0)$ on $\mathbb{P}^n$. Then

$$\check{\delta}_{\xi_0}(f) = \int_{\xi_0} f(x)\,dm(x)$$

so

(50) $$\check{\delta}_{\xi_0} = m_{\xi_0} ,$$

the Euclidean measure of ξ_0.

c) Let χ_B be the characteristic function of the unit ball $B \subset \mathbb{R}^n$. Then by (47),

$$\widehat{\chi}_B(\omega,p) = \begin{cases} \dfrac{\Omega_{n-1}}{n-1}(1-p^2)^{\frac{1}{2}(n-1)} & |p| \leq 1 \\ 0 & |p| > 1 \end{cases}$$

d) Let Ω be a bounded convex region in $\mathbb{R}^n$ whose boundary is a smooth surface. We shall obtain a formula for the volume of Ω in terms of the areas of its hyperplane sections. For simplicity we assume n odd. The characteristic function χ_Ω is a distribution of compact support and $(\chi_\Omega)\widehat{}$ is thus well defined. Approximating χ_Ω in the L^2 - norm by a sequence $(\psi_n) \subset C_c^\infty(\Omega)$ we see from Theorem 4.1 that $\partial_p^{\frac{1}{2}(n-1)}\widehat{\psi}_n(\omega,p)$ converges in the L^2 - norm on $\mathbb{P}^n$. Since

$$\int \widehat{\psi}_n(\xi)\phi(\xi)\,d\xi = \int \psi_n(x)\check{\phi}(x)\,dx$$

it follows from Schwarz' inequality that $\widehat{\psi}_n \longrightarrow (\chi_\Omega)\widehat{}$ in the sense of distributions and accordingly $\partial^{\frac{1}{2}(n-1)}\widehat{\psi}_n$ converges as a distribution to $\partial^{\frac{1}{2}(n-1)}((\chi_\Omega)\widehat{})$. Since the L^2 limit is also a limit in the sense of distributions this last function equals the L^2 limit of the sequence $\partial^{\frac{1}{2}(n-1)}\hat{\psi}_n$. From Theorem 4.1 we can thus conclude the following result:

THEOREM 5.8. Let $\Omega \subset \mathbb{R}^n$ (n odd) be a convex region as above and $V(\Omega)$ its volume. Let $A(\omega,p)$ denote the (n-1) - dimensional area of the intersection of Ω with the hyperplane $(x,\omega) = p$. Then

$$V(\Omega) = \frac{1}{2}(2\pi)^{1-n}\int_{\mathbb{S}^{n-1}}\int_{\mathbb{R}}\left|\frac{\partial^{\frac{1}{2}(n-1)}A(\omega,p)}{\partial p^{\frac{1}{2}(n-1)}}\right|^2 dp\,d\omega. \tag{51}$$

§6. Integration over d-planes. X-ray Transforms.

Let d be a fixed integer in the range $0 < d < n$. Since a hyperplane in $\mathbb{R}^n$ can be viewed as a disjoint union of parallel d-planes, parametrized by $\mathbb{R}^{n-1-d}$, it is obvious from (4) that if $f \in \mathcal{S}(\mathbb{R}^n)$ has 0 integral over each d-plane in $\mathbb{R}^n$ then it is identically 0. Similarly we can deduce the following consequence of Theorem 2.6.

COROLLARY 6.1. Let $f \in C(\mathbb{R}^n)$ satisfy the following conditions:

(i) For each integer $m>0$, $x^m f(x)$ is bounded on $\mathbb{R}^n$.

(ii) For each d-plane ξ_d outside the unit ball $|x|<1$ we have

$$\int_{\xi_d} f(x)\, dm(x) = 0$$

dm being the Euclidean measure.

Then

$$f(x) = 0 \quad \textit{for} \quad |x| > 1.$$

We define the *d-dimensional Radon transform* $f \to \hat{f}$ by

$$\hat{f}(\xi) = \int_{\xi} f(x)\,dm(x)\,, \quad \xi \text{ a d-plane.} \tag{52}$$

Because of the applications to radiology indicated in §7,b) the one-dimensional Radon transform is often called the *X-ray transform*. We can then reformulate Cor. 6.1 as follows.

COROLLARY 6.2. *Let* $f,g \in C(\mathbf{R}^n)$ *satisfy the rapid decrease condition: For each* $m > 0$, $|x|^m f(x)$ *and* $|x|^m g(x)$ *are bounded on* $\mathbf{R}^n$. *Assume for the d-dimensional Radon transforms*

$$\hat{f}(\xi) = \hat{g}(\xi)$$

whenever the d-plane ξ *lies outside the unit ball. Then*

$$f(x) = g(x) \quad \textit{for} \quad |x| > 1.$$

We shall now generalize the inversion formula in Theorem 3.1. If ϕ is a continuous function on the space of d-planes in $\mathbf{R}^n$ we denote by $\check{\phi}$ the point function

$$\check{\phi}(x) = \int_{x \in \xi} \phi(\xi)\,d\mu(\xi),$$

where μ is the unique measure on the (compact) space of d-planes passing through x, invariant under all rotations around x and with total measure 1. If σ is a fixed d-plane through the origin we have in analogy with (16),

$$\check{\phi}(x) = \int_K \phi(x + k\cdot\sigma)\,dk. \tag{53}$$

THEOREM 6.3. *The d-dimensional Radon transform in* $\mathbb{R}^n$ *is inverted by the formula*

$$cf = L^{\frac{1}{2}d}((\hat{f})^{\vee}), \qquad f \in \mathcal{S}(\mathbb{R}^n), \tag{54}$$

where

$$c = \frac{\Gamma(\frac{1}{2}n)}{\Gamma(\frac{1}{2}(n-d))}(-4\pi)^{\frac{1}{2}d}.$$

Proof. We have in analogy with (32)

$$(\hat{f})^{\vee}(x) = \int_K \left\{\int_\sigma f(x + k\cdot y)\,dm(y)\right\}dk$$

$$= \int_\sigma dm(y)\int_K f(x + k\cdot y)\,dk = \int_\sigma (M^{|y|}f)(x)\,dm(y).$$

Hence

$$(\hat{f})^{\vee}(x) = \Omega_d\int_0^\infty (M^r f)(x)\,r^{d-1}\,dr$$

so using polar coordinates around x,

$$(\hat{f})^{\vee}(x) = \frac{\Omega_d}{\Omega_n}\int_{\mathbb{R}^n} |x-y|^{d-n} f(y)\,dy. \tag{55}$$

The theorem now follows from Prop 8.6.

As a corrollary of Theorem 2.10 we now obtain a generalization, characterizing the image of the space $C_c^\infty(\mathbb{R}^n)$ under the d-dimensional Radon transform.

The set $\mathbb{G}(d,n)$ of d-planes in $\mathbb{R}^n$ is a manifold, in fact a homogeneous space of the group $M(n)$ of all isometries of $\mathbb{R}^n$. Let $\mathbb{G}_{d,n}$ denote the manifold of all d-dimensional subspaces (d-planes through 0) of $\mathbb{R}^n$. The parallel translation of a d-plane to one through 0 gives a mapping π of $\mathbb{G}(d,n)$ onto $\mathbb{G}_{d,n}$. The inverse image $\pi^{-1}(\sigma)$ of a member $\sigma \in \mathbb{G}_{d,n}$ is naturally identified with the orthogonal complement $\sigma^\perp$. Let us write $\xi = (\sigma,x'')$ if $\sigma = \pi(\xi)$ and $x'' = \sigma^\perp \cap \xi$. Then (52) can be written

$$\hat{f}(\sigma,x'') = \int_\sigma f(x' + x'')\,dx'. \tag{56}$$

For $k \in \mathbb{Z}^+$ we consider the polynomial

$$P_k(u) = \int_{\mathbb{R}^n} f(x)\,(x,u)^k\,dx. \tag{57}$$

If $u = u'' \in \sigma^\perp$ this can be written

$$\int_{\mathbb{R}^n} f(x)\,(x,u'')^k\,dx = \int_{\sigma^\perp}\int_\sigma f(x' + x'')\,(x'',u'')^k\,dx'\,dx''$$

so the polynomial

$$P_{\sigma,k}(u'') = \int_{\sigma^\perp} \hat{f}(\sigma,x'')\,(x'',u'')^k\,dx''$$

is the restriction to $\sigma^\perp$ of the polynomial P_k.

In analogy with the space $\mathcal{D}_H(\mathbb{P}^n)$ in No.2 we define the space $\mathcal{D}_H(\mathbb{G}(d,n))$ as the set of C^∞ functions $\phi(\xi) = \phi_\sigma(x'')$ on $\mathbb{G}(d,n)$ of compact support satisfying the following condition.

(H): *For each* $k \in \mathbb{Z}^+$ *there exists a homogeneous* k^{th} *degree polynomial* P_k *on* $\mathbb{R}^n$ *such that for each* $\sigma \in \mathbb{G}_{d,n}$ *the polynomial*

$$P_{\sigma,k}(u'') = \int_{\sigma^\perp} \phi_\sigma(x'')(x'',u'')^k dx'', \qquad u'' \in \sigma^\perp,$$

coincides with the restriction $P_k|\sigma$.

COROLLARY 6.4. *The d-dimensional Radon transform is a bijection of* $\mathcal{D}(\mathbb{R}^n)$ *onto* $\mathcal{D}_H(\mathbb{G}(d,n))$.

Proof. For $d = n - 1$ this is Theorem 2.10. We shall now reduce the case of general d to the case $d = n - 1$. It remains just to prove the surjectivity in Cor. 6.4.

Let $\phi \in \mathcal{D}_H(\mathbb{G}(d,n))$. Let $\omega \in \mathbb{R}^n$ be a unit vector. Choose a d-dimensional subspace σ perpendicular to ω and consider the (n-d-1)-dimensional integral

$$\Psi_\sigma(\omega,p) = \int_{(\omega,x'')=p,\, x''\in\sigma^\perp} \phi_\sigma(x'')d_{n-d-1}(x'') \qquad p \in \mathbb{R}. \tag{58}$$

We claim that this is independent of the choice of σ. In fact

$$\int_{\mathbb{R}} \Psi_\sigma(\omega,p)p^k dp = \int_{\mathbb{R}} p^k \left(\int \phi_\sigma(x'')d_{n-d-1}(x'')\right) dp$$

$$= \int_{\sigma^\perp} \phi_\sigma(x'')(x'',\omega)^k dx'' = P_k(\omega).$$

If we had chosen another σ, say σ_1, perpendicular to ω ,then by the above $\Psi_\sigma(\omega,p) - \Psi_{\sigma_1}(\omega,p)$ would have been orthogonal to all polynomials in p ; having compact support it would have been identically 0. Thus we have a well defined function $\Psi(\omega,p) = \Psi_\sigma(\omega,p)$ to which Theorem 2.10 applies. From this theorem we get a function $f \in \mathcal{D}(\mathbb{R}^n)$ such that

$$\Psi(\omega,p) = \int_{(x,\omega)=p} f(x)\,dm(x). \tag{59}$$

It remains to prove that

$$\phi_\sigma(x'') = \int_\sigma f(x' + x'')dx'. \tag{60}$$

But as x'' runs through an arbitrary hyperplane in $\sigma^\perp$ it follows from (58) and (59) that both sides of (60) have the same integral. By the injectivity of the (n-d-1) - dimensional Radon transform on $\sigma^\perp$, equation (60) follows. This proves Corollary 6.4.

§7. Applications.

a) Partial differential equations.

The inversion formula in Theorem 3.1 is very well suited for applications to partial differential equations. To explain the underlying principle we write the inversion formula in the form

$$f(x) = \gamma L_x^{\frac{n-1}{2}} \left(\int_{\mathbf{S}^{n-1}} \hat{f}(\omega, (x,\omega))\,d\omega \right), \tag{61}$$

where the constant γ equals $\frac{1}{2}(2\pi i)^{1-n}$. Note that the function $f_\omega(x) = \hat{f}(\omega,(x,\omega))$ is a <u>plane wave with normal</u> ω, that is, it is constant on each hyperplane perpendicular to ω.

Consider now a differential operator

$$D = \sum_{(k)} a_{k_1 \cdots k_n} \partial_1^{k_1} \cdots \partial_n^{k_n}$$

with constant coefficients $a_{k_1 \cdots k_n}$, and suppose we want to solve the differential equation

$$\text{(62)} \qquad Du = f$$

where f is a given function in $\mathcal{S}(\mathbb{R}^n)$. To simplify the use of (61) we assume n to be odd. We begin by considering the differential equation

$$\text{(63)} \qquad Dv = f_\omega,$$

where f_ω is the plane wave defined above and we look for a solution v which is also a plane wave with normal ω. But a plane wave with normal ω is just a function of one variable; also if v is a plane wave with normal ω so is the function Dv. The differential equation (63) (with v a plane wave) is therefore an <u>ordinary</u> differential equation with constant coefficients. Suppose $v = u_\omega$ is a solution and assume that this choice can be made smoothly in ω. Then the function

(64) $$u = \gamma L^{\frac{n-1}{2}} \int_{\mathbf{S}^{n-1}} u_\omega \, d\omega$$

is a solution to the differential equation (62). In fact, since D and $L^{\frac{n-1}{2}}$ commute we have

$$Du = \gamma L^{\frac{n-1}{2}} \int_{\mathbf{S}^{n-1}} Du_\omega \, d\omega = \gamma L^{\frac{n-1}{2}} \int_{\mathbf{S}^{n-1}} f_\omega \, d\omega = f.$$

This method only assumes that the plane wave solution u_ω to the ordinary differential equation $Dv = f_\omega$ exists and can be chosen so as to depend smoothly on ω. This cannot always be done because D might annihilate all plane waves with normal ω. (For example, take $D = \partial^2/\partial x_1 \partial x_2$ and $\omega = (1,0)$). However, if this restriction to plane waves is never 0 it follows from a theorem of Trèves [1963] that the solution u_ω can be chosen depending smoothly on ω. Thus we can state

THEOREM 7.1. Assuming the restriction D_ω of D to the space of plane waves with normal ω is $\neq 0$ for each ω formula (64) gives a solution to the differential equation $Du = f$ $(f \in \mathcal{S}(\mathbb{R}^n))$.

The method of plane waves can also be used to solve the Cauchy problem for hyperbolic differential equations with constant coefficients. We illustrate the method by means of the wave equation in $\mathbb{R}^n$,

(65) $$Lu = \frac{\partial^2 u}{\partial t^2} \qquad u(x,0) = 0, \quad \frac{\partial u}{\partial t}(x,0) = f(x),$$

$f \in \mathcal{S}(\mathbb{R}^n)$ being a given function.

LEMMA 7.2. For each $x \in \mathbb{R}^n$

$$\int_{\mathbb{S}^{n-1}} |(\omega,x)|^k d\omega = \frac{2\pi^{\frac{1}{2}(n-1)}\Gamma(\frac{1}{2}(k+1))}{\Gamma(\frac{1}{2}(k+n))} |x|^k . \tag{66}$$

Proof. The hyperplane perpendicular to x with distance p from 0 intersects $\mathbb{S}^{n-1}$ in an $(n-2)$ - sphere of radius $(1-p^2)^{\frac{1}{2}}$. The integrand is the constant $|x|p$ on this sphere and from the picture we have

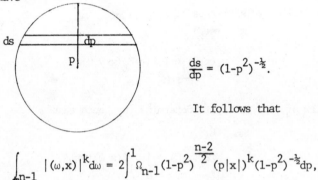

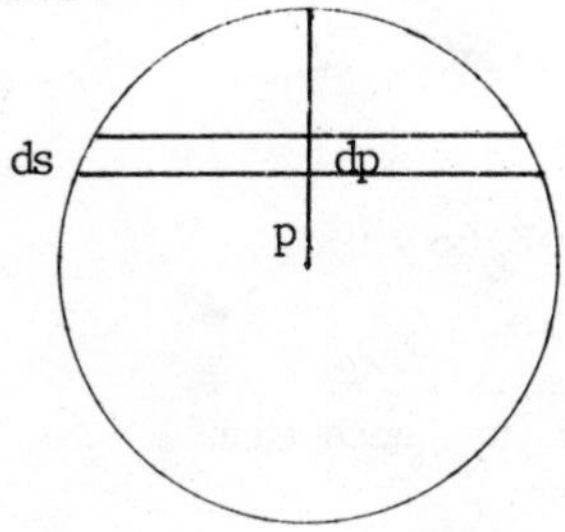

$$\frac{ds}{dp} = (1-p^2)^{-\frac{1}{2}}.$$

It follows that

$$\int_{\mathbb{S}^{n-1}} |(\omega,x)|^k d\omega = 2\int_0^1 \Omega_{n-1}(1-p^2)^{\frac{n-2}{2}}(p|x|)^k(1-p^2)^{-\frac{1}{2}}dp,$$

which implies the lemma.

Using (66) and the identities (88) and (89) in §8 we can derive the identity

$$L_x^{\frac{n+1}{2}} \int_{\mathbb{R}^n}\left(\int_{\mathbb{S}^{n-1}} f(y)\,|((x-y),\omega)|\,d\omega\right)dy = 4(2\pi i)^{n-1}f(x), \tag{67}$$

for n odd.

Suppose u_ω is a plane wave with normal ω satisfying the wave equation

$$Lu_\omega = \frac{\partial^2}{\partial t^2} u_\omega$$

ignoring the initial condition. Then the function

$$Z(x,t) = \int_{\mathbf{S}^{n-1}} u_\omega(x,t)\,d\omega$$

is also a solution and so is the convolution

$$v(x,t) = \int_{\mathbf{R}^n} f(y)Z(x-y,t)\,dy.$$

The time derivative satisfies

$$v_t(x,0) = \int_{\mathbf{R}^n} f(y)Z_t(x-y,0)\,dy$$

so in view of (67) we try to determine u_ω such that

$$Z_t(x,0) = \int_{\mathbf{S}^{n-1}} |(x,\omega)|\,d\omega\,.$$

For this we put

$$u_\omega(x,t) = \tfrac{1}{2}((x,\omega) + t)^2\operatorname{sgn}((x,\omega) + t),$$

and have then obtained the following result.

THEOREM 7.3. _Let_ n _be odd. A solution to the Cauchy problem_ (65) _is given by_

$$u(x,t) = \frac{1}{8(2\pi i)^{n-1}} L_x^{\frac{n+1}{2}} \left[\int_{\mathbf{R}^n} f(x-y) \int_{\mathbf{S}^{n-1}} ((y,\omega)+t)^2\operatorname{sgn}((y,\omega)+t)\,d\omega\,dy \right].$$

We shall now put this into another form for $n = 3$. Since u_ω satisfies the wave equation the formula can be written

$$u(x,t) = -\frac{1}{16\pi^2}\frac{\partial}{\partial t}\int_{R^3} f(y)\left[\frac{\partial}{\partial t}\int_{\mathbb{S}^2} \operatorname{sgn}\{(x-y,\omega) + t\}d\omega\right] dy.$$

As for Lemma 7.2 we compute the $\mathbb{S}^2$ integral by slices perpendicular to x-y. This gives

$$\int_{\mathbb{S}^2} \operatorname{sgn}\{(x-y,\omega) + t\}d\omega = 2\pi\int_{-1}^{1} \operatorname{sgn}\{|x-y|p + t\}dp,$$

which, since $t \geqslant 0$, equals

$$2\pi\int_{-1}^{1} \operatorname{sgn}\{p + |x-y|^{-1}t\}dp = \begin{cases} 4\pi|x-y|^{-1}t & \text{if} \quad t < |x-y| \\ 4\pi & \text{if} \quad t > |x-y|. \end{cases}$$

Hence

$$\begin{aligned} u(x,t) &= -\frac{1}{4\pi}\frac{\partial}{\partial t}\int_{|x-y|>t} f(y)|x-y|^{-1}dy \\ &= -\frac{1}{4\pi}\frac{\partial}{\partial t}\int_t^\infty d\rho\int_{S^\rho(x)} \rho^{-1}f(y)d\omega(y) \\ &= \frac{1}{4\pi}(4\pi t^2)t^{-1}(M^t f)(x). \end{aligned}$$

This proves the following result.

COROLLARY 7.4. *For* $n = 3$ *the solution to the Cauchy problem* (65) *is given by*

$$u(x,t) = t(M^t f)(x).$$

REMARKS. (i) A similar formula holds for arbitrary n : the solution to (65) is given by

$$u(x,t) = \frac{1}{(n-2)!}\frac{\partial^{n-2}}{\partial t^{n-2}}\int_0^t (M^\rho f)(x)\rho(t^2-\rho^2)^{\frac{1}{2}(n-3)}d\rho.$$

(cf. John [1955], Ch. II).

(ii) (Huygens' principle). The formula in Cor 7.4 shows that $u(x,t)$ is determined by the values of f on the sphere $S^t(x)$ in $\mathbf{R}^3$. This phenomenon, called Huygens' principle, remains true for $\mathbf{R}^n$ (n odd) in a slightly weaker form. In fact, if the differentiations $\frac{\partial^{n-2}}{\partial t^{n-2}}$ are carried out we end up with a linear combination with polynomial coefficients,

$$\sum_k P_k(t)\frac{\partial^k(M^t f)(x)}{\partial t^k},$$

which shows the following:

For n odd, the solution $u(x,t)$ to the Cauchy problem (65) is determined by the values of f in an arbitrarily thin shell around $S^t(x)$.

For even n the integral will remain so $u(x,t)$ will require knowing f in the ball $B^t(x)$.

b) X-ray Reconstruction.

The classical interpretation of an X-ray picture is an attempt at reconstructing properties of a three-dimensional body by means of the X-ray projection on a plane.

In modern X-ray technology the picture is given a more refined mathematical interpretation. Let $B \subset \mathbf{R}^3$ be a body (for example a part of a human body) and let $f(x)$ denote its density at a point x. Let ξ be a line in $\mathbf{R}^3$ and suppose a thin beam of X-rays is directed at B along ξ. Let I_0 and I respectively, denote the intensity of the beam before entering B and after leaving B. It is then a physically accepted fact that

$$\log(I_0/I) = \int_\xi f(x)\,dm(x), \tag{68}$$

the integral $\hat{f}(\xi)$ of f along ξ. Since the left hand side is determined by the X-ray picture, the <u>X-ray reconstruction problem</u> amounts to the determination of the function f by means of its line integrals $\hat{f}(\xi)$. <u>The inversion formula in Theorem 2.27 gives an explicit solution of this problem</u>.

If $B_0 \subset B$ is a convex subset (for example the heart) it may be of interest to determine the density f outside B_0 using only X-rays which do not intersect B_0. <u>The support theorem (Theorem 2.6, Cor. 2.8 and 6.2) implies that</u> f <u>is determined outside</u> B_0 <u>on the basis of the integrals</u> $\hat{f}(\xi)$ <u>for which</u> ξ <u>does not intersect</u> B_0.

In practice one can of course only determine the integrals $\hat{f}(\xi)$ in (68) for <u>finitely</u> many directions. A compensation for this is the fact that only an approximation to the density f is required. One then encounters the mathematical problem of selecting the directions so as to optimize the approximation.

As before we represent the line ξ as the pair $\xi = (\omega, z)$ where $\omega \in \mathbb{R}^n$ is a unit vector in the direction of ξ and $z = \xi \cap \omega^\perp$ ($\perp$ denoting orthogonal complement). We then write

$$(69) \qquad \hat{f}(\xi) = \hat{f}(\omega, z) = (P_\omega f)(z).$$

The function $P_\omega f$ is the X-ray picture or the radiograph in the direction ω. Here f is a function on $\mathbb{R}^n$ vanishing outside a ball B around the origin and for the sake of Hilbert space methods to be used it is convenient to assume in addition that $f \in L^2(B)$. Then $f \in L^1(\mathbb{R}^n)$ so by the Fubini theorem we have: for each $\omega \in \mathbb{S}^{n-1}$, $P_\omega f(z)$ is defined for almost all $z \in \omega^\perp$. Moreover, we have in analogy with (4),

$$(70) \qquad \tilde{f}(\zeta) = \int_{\omega^\perp} (P_\omega f)(z) e^{-i(z,\zeta)} dz, \qquad (\zeta \in \omega^\perp).$$

PROPOSITION 7.5. An object is determined by any infinite set of radiographs.

In other words, a compactly supported function f is determined by the functions $P_\omega f$ for any infinite set of ω.

Proof. Since f has compact support $\tilde{f}$ is an analytic function on $\mathbb{R}^n$. But if $\tilde{f}(\zeta) = 0$ for $\zeta \in \omega^\perp$ we have
$\tilde{f}(\eta) = (\omega, \eta) g(\eta) \quad (\eta \in \mathbb{R}^n)$ where g is also analytic. If
$P_{\omega_1} f, \ldots, P_{\omega_k} f \ldots$ all vanish identically for an infinite set
$\omega_1, \ldots, \omega_k \ldots$ we see that for each k

$$\tilde{f}(\eta) = \prod_{i=1}^{k} (\omega_i, \eta) g_k(\eta),$$

where g_k is analytic. But this would contradict the power series expansion of $\tilde{f}$ which shows that for a suitable $\omega \in \mathbf{S}^{n-1}$ and integer $r \geq 0$, $\lim_{t \to 0} \tilde{f}(t\omega)t^{-r} \neq 0$.

If only finitely many radiographs are used we get the opposite result.

PROPOSITION 7.6. Let $\omega_1, \ldots, \omega_k \in \mathbf{S}^{n-1}$ <u>be an arbitrary finite set</u>. <u>Then there exists a function</u> $f \in C_c^\infty(\mathbf{R}^n)$, $f \not\equiv 0$ <u>such that</u>

$$P_{\omega_i} f \equiv 0 \qquad \underline{\text{for all}} \qquad 1 \leq i \leq k.$$

<u>Proof</u>. We have to find $f \in C_c^\infty(\mathbf{R}^n)$, $f \not\equiv 0$, such that $\tilde{f}(\zeta) = 0$ for $\zeta \in \omega_i^\perp (1 \leq i \leq k)$. For this let D be the constant coefficient differential operator such that

$$(Du)^\sim(\eta) = \prod_1^k (\omega_i, \eta)\tilde{u}(\eta), \qquad \eta \in \mathbf{R}^n.$$

If $u \not\equiv 0$ is any function in $C_c^\infty(\mathbf{R}^n)$ then $f = Du$ has the desired property.

We next consider the problem of <u>approximate reconstruction</u> of the function f from a finite set of radiographs $P_{\omega_1} f, \ldots, P_{\omega_k} f$.

Let N_j denote the null space of P_{ω_j} and let P_j the orthogonal projection of $L^2(B)$ on the plane $f + N_j$; in other words

(71) $$P_j g = Q_j(g-f) + f,$$

where Q_j is the (linear) projection onto the subspace $N_j \subset L^2(B)$. Put $P = P_k \dots P_1$. Let $g \in L^2(B)$ be arbitrary (the initial guess for f) and form the sequence $P^m g$, $m = 1,2,\dots$. Let $N_o = \bigcap_1^k N_j$ and let P_o (resp. Q_o) denote the orthogonal projection of $L^2(B)$ on the plane $f + N_o$ (subspace N_o). We shall prove that the sequence $P^m g$ converges to the projection $P_o g$. This is natural since by $P_o g - f \in N_o$, $P_o g$ and f have the same radiographs in the directions $\omega_1, \dots, \omega_k$.

THEOREM 7.7. *With the notations above,*

$$P^m g \longrightarrow P_o g \quad \underline{\text{as}} \quad m \longrightarrow \infty$$

for each $g \in L^2(B)$.

Proof. We have, by iteration of (71)

$$(P_k \dots P_1)g - f = (Q_k \dots Q_1)(g - f)$$

and, putting $Q = Q_k \dots Q_1$ we obtain

$$P^m g - f = Q^m(g - f).$$

We shall now prove that $Q^m g \longrightarrow Q_o g$ for each g; since $P_o g = Q_o(g - f) + f$ this would prove the result. But the statement about Q^m comes from the following general result about abstract Hilbert space.

THEOREM 7.8. *Let* $\mathcal{H}$ *be a Hilbert space and* Q_i *the projection of* $\mathcal{H}$ *onto a subspace* $N_i \subset \mathcal{H}$ $(1 \leqslant i \leqslant k)$. *Let* $N_o = \bigcap_1^k N_i$ *and* $Q_o: \mathcal{H} \longrightarrow N_o$ *the projection. Then if* $Q = Q_k \cdots Q_1$

$$Q^m g \longrightarrow Q_o g \qquad \textit{for each} \qquad g \in \mathcal{H}.$$

Since Q is a contraction $(\|Q\| \leqslant 1)$ we begin by proving a simple lemma about such operators.

LEMMA 7.9. *Let* $T: \mathcal{H} \longrightarrow \mathcal{H}$ *be a linear operator of norm* $\leqslant 1$. *Then*

$$\mathcal{H} = C\ell((I - T)\mathcal{H}) \oplus \text{Null space } (I - T)$$

is an orthogonal decomposition, $C\ell$ *denoting closure, and* I *the identity.*

Proof. If $Tg = g$ then since $\|T^*\| = \|T\| \leqslant 1$ we have $\|g\|^2 = (g,g) = (Tg,g) = (g,T^*g) \leqslant \|g\| \, \|T^*g\| \leqslant \|g\|^2$ so all terms in the inequalities are equal. Hence

$$\|g - T^*g\|^2 = \|g\|^2 - (g,T^*g) - (T^*g,g) + \|T^*g\|^2 = 0$$

so $T^*g = g$. Thus $I - T$ and $I - T^*$ have the same null space. But $(I - T^*)g = 0$ is equivalent to $(g,(I - T)\mathcal{H}) = 0$ so the lemma follows.

Definition. An operator T on a Hilbert space $\mathcal{H}$ is said to have *property (S)* if

$$\|f_n\| \leqslant 1, \|Tf_n\| \longrightarrow 1 \quad \text{implies} \quad \|(I - T)f_n\| \longrightarrow 0 . \tag{72}$$

LEMMA 7.10. *A projection, and more generally a finite product of projections, has property* (S).

Proof. If T is a projection then

$$\|(I - T)f_n\|^2 = \|f_n\|^2 - \|Tf_n\|^2 \leq 1 - \|Tf_n\|^2 \longrightarrow 0$$

whenever

$$\|f_n\| \leq 1 \quad \text{and} \quad \|Tf_n\| \longrightarrow 1.$$

Let T_2 be a projection and suppose T_1 has property (S) and $\|T_1\| \leq 1$. Suppose $f_n \in \mathcal{H}$ and $\|f_n\| \leqslant 1$, $\|T_2T_1f_n\| \longrightarrow 1$. The inequality implies $\|T_1f_n\| \leqslant 1$ and since

$$\|T_1f_n\|^2 = \|T_2T_1f_n\|^2 + \|(I - T_2)(T_1f_n)\|^2$$

we also deduce $\|T_1f_n\| \longrightarrow 1$. Writing

$$(I - T_2T_1)f_n = (I - T_1)f_n + (I - T_2)T_1f_n$$

we conclude that T_2T_1 has property (S). The lemma now follows by induction.

LEMMA 7.11. Suppose T has property (S) and $\|T\| \leq 1$. Then for each $f \in \mathcal{H}$

$$T^n f \longrightarrow \pi f \qquad \text{as} \qquad n \longrightarrow \infty ,$$

where π is the projection onto the fixed point space of T.

Proof. Let $f \in \mathcal{H}$. Since $\|T\| \leqslant 1$, $\|T^n f\|$ decreases monotonically to a limit $\alpha \geqslant 0$. If $\alpha = 0$ we have $T^n f \longrightarrow 0$. By Lemma 7.9 $\pi T = T\pi$ so $\pi f = T^n \pi f = \pi T^n f$ so $\pi f = 0$ in this case. If $\alpha > 0$ we put $g_n = \|T^n f\|^{-1}(T^n f)$. Then $\|g_n\| = 1$ and $\|Tg_n\| \longrightarrow 1$. Since T has property (S) we deduce

$$T^n(I - T)f = \|T^n f\|(I - T)g_n \longrightarrow 0.$$

Thus $T^n h \longrightarrow 0$ for all h in the range of $I - T$. If g is in the closure of this range then given $\varepsilon > 0$ there exist $h \in (I - T)\mathcal{H}$ such that $\|g - h\| < \in$. Then

$$\|T^n g\| \leqslant \|T^n(g - h)\| + \|T^n h\| < \in + \|T^n h\|$$

whence $T^n g \longrightarrow 0$. On the other hand, if h is in the null space of $I - T$ then $Th = h$ so $T^n h \longrightarrow h$. Now the lemma follows from Lemma 7.9.

In order to deduce Theorem 7.8 from Lemmas 7.10 and 7.11 we just have to verify that N_0 is the fixed point space of Q. But if $Qg = g$ then

$$\|g\| = \|Q_k \cdots Q_1 g\| \leq \|Q_{k-1} \cdots Q_1 g\| \leq \cdots \leq \|Q_1 g\| \leq \|g\|$$

so equality signs hold everywhere. But the Q_i are projections so the norm identities imply

$$g = Q_1 g = Q_2 Q_1 g = \cdots = Q_k \cdots Q_1 g$$

which shows $g \in N_o$.

§8. Appendix. Distributions and Riesz Potentials.

In this section we develop the main results in the theory of Riesz potentials in $\mathbb{R}^n$([1949]). While most of the results are readily available in the literature, some are not, so we give here a self-contained exposition. First we recall briefly some notions from distribution theory.

If $I \subset \mathbb{R}$ is a compact interval let $C_I^\infty(\mathbb{R}) \subset C_c^\infty(\mathbb{R})$ denote the subspace of functions with support in I. This subspace has a topology given by the norms (6); in other words a sequence (f_n) in $C_I^\infty(\mathbb{R})$ converges to 0 if each derivative $f_n^{(h)}$ converges for $n \longrightarrow \infty$ to 0 uniformly on $\mathbb{R}$. A linear functional T on $C_c^\infty(\mathbb{R})$ is called a <u>distribution</u> on $\mathbb{R}$ its restriction to each $C_I^\infty(\mathbb{R})$ is continuous. Each locally integrable function F on $\mathbb{R}$ gives rise to a distribution $T_F : \phi \longrightarrow \int \phi(x)F(x)dx$ on $\mathbb{R}$. The <u>derivative</u> of a distribution T is the distribution $\phi \longrightarrow -T(\phi')$ on $\mathbb{R}$. If $F \in C^1(\mathbb{R})$ then the distributions $T_{F'}$ and $(T_F)'$ coincide (integration by parts).

A linear form on the space $\mathcal{S}(\mathbb{R})$ is called a <u>tempered distribution</u> if it is continuous in the topology given by the norms (6). The restriction of a tempered distribution to $C_c^\infty(\mathbb{R})$ is a distribution and since the subspace $C_c^\infty(\mathbb{R}) \subset \mathcal{S}(\mathbb{R})$ is dense two tempered distribution coincide if the coincide on $C_c^\infty(\mathbb{R})$.

If T is a tempered distribution on $\mathbb{R}$ its <u>Fourier transform</u> $\tilde{T}$ is the linear form on $\mathcal{S}(\mathbb{R})$ defined by

$$\tilde{T}(\phi) = T(\tilde{\phi}), \qquad \phi \in \mathcal{S}(\mathbb{R}),$$

where $\tilde{\phi}$ is the Fourier transform defined in §2. Since $\phi \longrightarrow \tilde{\phi}$ is a homeomorphism of $\mathcal{S}(\mathbb{R})$ onto itself it follows that $\tilde{T}$ is another tempered distribution. Since $\int F\tilde{\phi} = \int \tilde{F}\phi \quad (\phi, F \in \mathcal{S}(\mathbb{R}))$ the distributions $(T_F)^\sim$ and $T_{\tilde{F}}$ coincide.

Since distributions generalize measures it is sometimes convenient to write

$$T(\phi) = \int_{\mathbb{R}} \phi(x)\,dT(x)$$

for the value of a distribution T on the function ϕ. A distribution is said to be 0 on an open set $U \subset \mathbb{R}$ if $T(\phi) = 0$ for each $\phi \in C_c^\infty(\mathbb{R})$ with support contained in U. If U is the union of all open subsets $U_\alpha \subset \mathbb{R}$ on which T is 0 then a partition of unity argument shows that $T = 0$ on U. The complement of U is called the <u>support of</u> T. A distribution T of <u>compact support</u> extends to a linear form on $C^\infty(\mathbb{R})$ by putting

$$T(\phi) = T(\phi_0\phi) \qquad \phi \in C^\infty(\mathbb{R})$$

if ϕ_0 is any function in $C_c^\infty(\mathbb{R})$ which is identically 1 on the support of T. The choice of ϕ_0 is immaterial. Moreover, T is tempered.

If S and T are two distributions at least one of compact support their <u>convolution</u> is the distribution defined by

$$\phi \longrightarrow \int_{\mathbb{R}}\int_{\mathbb{R}} \phi(x+y)dS(x)dT(y) \ , \quad \phi \in C_c^\infty(\mathbb{R}) .$$

This distribution is denoted $S*T$. If $f \in C_c^\infty(\mathbb{R})$ the distribution $T_f * T$ has the form T_g where

$$g(x) = \int f(x-y)dT(y),$$

so we write for simplicity $g = f * T$.

If S and T both have compact support so does $S * T$. Also $\tilde{S}$ and $\tilde{T}$ have the form $\tilde{S} = T_s$, $\tilde{T} = T_t$ where $s,t \in C^\infty(\mathbb{R})$ and in addition $(S * T)^\sim = T_{st}$. We express this in the form

$$(S*T)^\sim = \tilde{S}\,\tilde{T} .$$

This formula is also true if $S \in \mathcal{S}(\mathbb{R})$ and if T is a tempered distribution.

These notions (distributions, tempered distributions, derivative, Fourier transform and convolution) generalize in an obvious manner to several variables. We often use Schwartz' notation $\mathcal{D}(\mathbb{R}^n)$ for $C_c^\infty(\mathbb{R})$, $\mathcal{E}(\mathbb{R}^n)$ for $C^\infty(\mathbb{R}^n)$ and $\mathcal{S}(\mathbb{R}^n)$ for the space of rapidly decreasing functions. The set of all distributions on $\mathbb{R}^n$ is denoted by $\mathcal{D}'(\mathbb{R}^n)$, the set of tempered distributions by $\mathcal{S}'(\mathbb{R}^n)$ and the set of compactly supported distributions by $\mathcal{E}'(\mathbb{R}^n)$. We have the canonical inclusions

$$\begin{array}{ccccc} \mathcal{D}(\mathbb{R}^n) & \subset & \mathcal{S}(\mathbb{R}^n) & \subset & \mathcal{E}(\mathbb{R}^n) \\ \cap & & \cap & & \cap \\ \mathcal{E}'(\mathbb{R}^n) & \subset & \mathcal{S}'(\mathbb{R}^n) & \subset & \mathcal{D}'(\mathbb{R}^n). \end{array}$$

We shall now study some useful examples in detail.

If $\alpha \in \mathbb{C}$ satisfies $\mathrm{Re}\,\alpha > -1$ the functional

$$x_+^\alpha : \phi \longrightarrow \int_0^\infty x^\alpha \phi(x)dx, \qquad \phi \in \mathcal{S}(\mathbb{R}), \tag{73}$$

is a well-defined tempered distribution. The mapping $\alpha \longrightarrow x_+^\alpha$ from the half plane $\mathrm{Re}\,\alpha > -1$ to the space $\mathcal{S}'(\mathbb{R})$ of tempered distributions is holomorphic (that is $\alpha \longrightarrow x_+^\alpha(\phi)$ is holomorphic for each $\phi \in \mathcal{S}(\mathbb{R})$). Writing

$$x_+^\alpha(\phi) = \int_0^1 x^\alpha(\phi(x) - \phi(o))dx + \frac{\phi(o)}{\alpha+1} + \int_1^\infty x^\alpha \phi(x)dx$$

the function $\alpha \longrightarrow x_+^\alpha$ is continued to a holomorphic function in the region $\mathrm{Re}\,\alpha > -2$, $\alpha \neq -1$. In fact

$$\phi(x) - \phi(0) = x\int_0^\infty \phi'(tx)dt,$$

so the first integral above converges for $\mathrm{Re}\,\alpha > -2$. More generally, $\alpha \longrightarrow x_+^\alpha$ can be extended to a holomorphic $\mathcal{S}'(\mathbb{R})$ - valued mapping in the region

$$\mathrm{Re}\,\alpha > -n-1, \quad \alpha \neq -1,-2,\ldots-n,$$

by means of the formula

$$x_+^\alpha(\phi) = \int_0^1 x^\alpha \,[\phi(x) - \phi(0) - x\phi'(0) - \ldots - \frac{x^{n-1}}{(n-1)!}\phi^{(n-1)}(0)]dx$$

$$(74) \qquad + \int_1^\infty x^\alpha \phi(x)dx + \sum_{k=1}^{n} \frac{\phi^{(k-1)}(0)}{(k-1)!(\alpha+k)} .$$

In this manner $\alpha \longrightarrow x_+^\alpha$ is a meromorphic distribution-valued function on $\mathbb{C}$, with simple poles at $\alpha = -1, -2, \ldots$. We note that the residue at $\alpha = -k$ is given by

$$(75) \qquad \operatorname*{Res}_{\alpha=-k} x_+^\alpha = \lim_{\alpha\to -k}(\alpha + k)x_+^\alpha = \frac{(-1)^{k-1}}{(k-1)!}\delta^{(k-1)} .$$

Here $\delta^{(h)}$ is the hth derivative of the delta-distribution δ. We note that x_+^α is always a tempered distribution.

Next we consider for $\operatorname{Re} \alpha > -n$ the distribution r^α on $\mathbb{R}^n$ given by

$$r^\alpha : \phi \longrightarrow \int_{\mathbb{R}^n} \phi(x)|x|^\alpha dx, \qquad \phi \in \mathcal{D}(\mathbb{R}^n).$$

LEMMA 8.1. *The mapping* $\alpha \longrightarrow r^\alpha$ *extends uniquely to a meromorphic mapping from* $\mathbb{C}$ *to the space* $\mathcal{S}'(\mathbb{R}^n)$ *of tempered distributions. The poles are the points*

$$\alpha = -n - 2h \qquad (h \in \mathbb{Z}^+)$$

and they are all simple.

Proof. We have for Re $\alpha > -n$

(76) $$r^{\alpha}(\phi) = \Omega_n \int_0^{\infty} (M^t\phi)(0)t^{\alpha+n-1}dt.$$

Next we note (say from (15) in §2) that the mean value function $t \longrightarrow (M^t\phi)(0)$ extends to an even C^{∞} function on $\mathbb{R}$, and its odd order derivatives at the origin vanish. Each even order derivative is non zero if ϕ is sutiably chosen. Since by (76)

(77) $$r^{\alpha}(\phi) = \Omega_n t_+^{\alpha+n-1}((M^t\phi)(0))$$

the first statement of the lemma follows. The possible (simple) poles of r^{α} are by the remarks about x_+^{α} given by $\alpha+n-1 = -1, -2, \ldots$. However if $\alpha+n-1 = -2, -4, \ldots$, formula (75) shows, since $(M^t\phi(0))^{(h)} = 0$, (h odd) that $r^{\alpha}(\phi)$ is holomorphic at the points $\alpha = -n-1, -n-3, \ldots$.

The remark about the even derivatives of $M^t\phi$ shows on the other hand, that the points $\alpha = -n-2h$ $(h \in \mathbb{Z}^+)$ are genuine poles. We note also from (75) and (77) that

(78) $$\operatorname*{Res}_{\alpha=-n} r^{\alpha} = \lim_{\alpha\to -n} (\alpha+n)r^{\alpha} = \Omega_n\delta .$$

We recall now that the Fourier transform $T \longrightarrow \tilde{T}$ of a tempered distribution T on $\mathbb{R}^n$ is defined by

$$\tilde{T}(\phi) = T(\tilde{\phi}) \qquad \phi \in \mathcal{S}(\mathbb{R}^n).$$

We shall now calculate the Fourier transforms of these tempered distributions r^α.

LEMMA 8.2. *We have the identities*

(79) $$(r^\alpha)^\sim = 2^{n+\alpha}\pi^{\frac{n}{2}}\frac{\Gamma(\frac{1}{2}(n+\alpha))}{\Gamma(-\frac{1}{2}\alpha)}\, r^{-\alpha-n}, \quad (\alpha,-\alpha-n\notin 2\,\mathbf{Z}^+)$$

(80) $$(r^{2h})^\sim = (2\pi)^n\,(-L)^h\delta, \qquad h\in\mathbf{Z}^+ .$$

Proof. We use the fact that if $\psi(x) = e^{-\frac{1}{2}|x|^2}$ then $\tilde\psi(u) = (2\pi)^{\frac{n}{2}}e^{-\frac{1}{2}|u|^2}$ so by the formula $\int f\tilde g = \int \tilde f g$ we obtain for $\phi\in\mathcal{S}(\mathbb{R}^n)$, $t>0$,

(81) $$\int\tilde\phi(x)e^{-\frac{1}{2}t|x|^2}dx = (2\pi)^{\frac{n}{2}}t^{-\frac{n}{2}}\int\phi(u)e^{-\frac{1}{2}t^{-1}|u|^2}du.$$

We multiply this equation by $t^{-\frac{1}{2}\alpha-1}$ and integrate with respect to t. On the left we obtain, using the formula

$$\int_0^\infty e^{-\frac{1}{2}t|x|^2}\,t^{-\frac{1}{2}\alpha-1}\,dt = \Gamma(-\tfrac{\alpha}{2})2^{-\frac{\alpha}{2}}\,|x|^\alpha$$

the expression

$$\Gamma(-\tfrac{\alpha}{2})\,2^{-\frac{\alpha}{2}}\int\tilde\phi(x)\,|x|^\alpha dx.$$

On the right we similarly obtain

$$(2\pi)^{\frac{n}{2}}\,\Gamma(\tfrac{n+\alpha}{2})2^{\frac{n+\alpha}{2}}\int\phi(u)\,|u|^{-\alpha-n}du.$$

The interchange of the integrations is valid for α in the strip $-n < \operatorname{Re}\alpha < 0$ so (79) is proved for these α. For the remaining ones it follows by analytic continuation. Finally, (80) is immediate from the definitions.

LEMMA 8.3. <u>The action of the Laplacian is given by</u>

(82) $$Lr^{\alpha} = \alpha(\alpha + n-2)r^{\alpha-2}, \qquad (-\alpha-n+2 \notin 2\mathbb{Z}^{+})$$

(83) $$Lr^{2-n} = (2-n)\Omega_n\delta \qquad (n \neq 2).$$

<u>For</u> $n = 2$ <u>this 'Poisson equation' is replaced by</u>

(84) $$L(\log r) = 2\pi\delta.$$

<u>Proof</u>. For $\operatorname{Re}\alpha$ sufficiently large (82) is obvious by computation. For the remaining ones it follows by analytic continuation. For (83) we use the Fourier transform and the fact that for a tempered distribution S,

$$(-LS)^{\sim} = r^2\tilde{S}.$$

Hence, by (79),

$$(-Lr^{2-n})^{\sim} = 4\,\frac{\pi^{\frac{n}{2}}}{\Gamma(\frac{n}{2}-1)} = \frac{2\pi^{\frac{n}{2}}}{\Gamma(\frac{n}{2})}\,(n-2)\tilde{\delta}.$$

Finally, we prove (84). If $\phi \in \mathcal{D}(\mathbb{R}^2)$ we have, putting

$F(r) = (M^r\phi)(0)$,

$$(L(\log r))(\phi) = \int_{\mathbb{R}^2} \log r\,(L\phi)(x)\,dx = \int_0^\infty (\log r)\,2\pi r(M^r L\phi)(0)\,dr\,.$$

Using Lemma 3.2 this becomes

$$\int_0^\infty \log r\;\; 2\pi r\{F''(r) + r^{-1}F'(r)\}dr,$$

which by integration by parts reduces to

$$\Big[\log r\,(2\pi r)F'(r)\Big]_0^\infty - 2\pi\int_0^\infty F'(r)dr = 2\pi F(0).$$

This proves (84).

<u>Méthode[1] erroné</u>. Another method for (84) is to put $n = 2$ in (82) and differentiate with respect to α. Since $\frac{d}{d\alpha}r^\alpha = \log r\, r^\alpha$ (by analytic continuation) we obtain

$$L(\log r\;\; r^\alpha) = 2\alpha r^{\alpha-2} + \alpha^2 \log r\, r^{\alpha-2}, \quad (-\alpha \notin 2\,\mathbb{Z}^+).$$

From (78) we have

$$\lim_{\alpha\to 0} \alpha r^{\alpha-2} = \lim_{\beta\to -2} (\beta + 2)r^\beta = 2\pi\delta$$

and $\lim_{\alpha\to 0} \alpha^2 r^{\alpha-2} = 0$ so we would appear to get the wrong result $L(\log r) = 4\pi\delta$. As pointed out to me by R. Melrose the source of the

1) Cf. M. Riesz [1949], p. 74.

error is the fact that multiplication by $\log r$ is not a continuous map of $\mathcal{S}'$ into $\mathcal{S}'$. A way out is to write (82) in the form $L(\alpha^{-1}(r^{\alpha}-1)) = \alpha r^{\alpha-2}$. Then (84) follows by letting $\alpha \longrightarrow 0$.

We shall now define fractional powers of L, motivated by the formula

$$(-Lf)^{\sim}(u) = |u|^2\tilde{f}(u),$$

so that formally we should like to have a relation

$$((-L)^p f)^{\sim}(u) = |u|^{2p}\tilde{f}(u).$$

Since the Fourier transform of a convolution is the product of the Fourier transforms, formula (79) (for $2p = -\alpha-n$) suggests defining

(85) $$(-L)^p f = I^{-2p}(f),$$

where I^{γ} is the <u>Riesz potential</u>

(86) $$(I^{\gamma}f)(x) = \frac{1}{H_n(\gamma)}\int_{\mathbb{R}^n} f(y)|x-y|^{\gamma-n}dy$$

with

(87) $$H_n(\gamma) = 2^{\gamma}\pi^{\frac{n}{2}}\frac{\Gamma(\frac{\gamma}{2})}{\Gamma(\frac{n-\gamma}{2})}.$$

Writing (86) as $H_n(\gamma)^{-1}(f * r^{\gamma-n})(x)$ and assuming $f \in \mathcal{S}(\mathbb{R}^n)$ we see that the poles of $r^{\gamma-n}$ are cancelled by the poles of $\Gamma(\frac{1}{2}\gamma)$ so $(I^{\gamma}f)(x)$ extends to a holomorphic function in $\mathbb{C}_n = \{\gamma\in\mathbb{C} : \gamma-n \notin 2\mathbb{Z}^+\}$. We have also by (78) and the formula for Ω_n

$$I^{0}f = \lim_{\gamma\to 0} I^{\gamma}f = f. \tag{88}$$

Furthermore, by (82) and analytic continuation,

$$I^{\gamma}Lf = LI^{\gamma}f = -I^{\gamma-2}f, \qquad f\in\mathcal{S}(\mathbb{R}^{n}),\ \gamma\in\mathbb{C}_{n}. \tag{89}$$

We now prove an important property of the Riesz' potentials.

PROPOSITION 8.4. _The following identity holds_:

$$I^{\alpha}(I^{\beta}f) = I^{\alpha+\beta}f \quad \text{for} \quad f\in\mathcal{S}(\mathbb{R}^{n}),\ \mathrm{Re}\,\alpha,\ \mathrm{Re}\,\beta > 0,\ \mathrm{Re}(\alpha+\beta) < n.$$

Proof. We have

$$I^{\alpha}(I^{\beta}f)(x) = \frac{1}{H_n(\alpha)}\int |x-z|^{\alpha-n}\left(\frac{1}{H_n(\beta)}\int f(y)\,|z-y|^{\beta-n}dy\right)dz$$

$$= \frac{1}{H_n(\alpha)H_n(\beta)}\int f(y)\left(\int |x-z|^{\alpha-n}|z-y|^{\beta-n}dz\right)dy.$$

The substitution $v = (x-z)\ /\ |x-y|$ reduces the inner integral to the form

$$|x-y|^{\alpha+\beta-n}\int_{\mathbb{R}^{n}} |v|^{\alpha-n}\ |w-v|^{\beta-n}dv, \tag{90}$$

where w is the unit vector $(x-y)\ /\ |x-y|$. Using a rotation around the origin we see that the integral in (90) equals the number

(91) $$c_n(\alpha,\beta) = \int_{\mathbb{R}^n} |v|^{\alpha-n}|e_1-v|^{\beta-n}dv,$$

where $e_1 = (1,0,\ldots,0)$. The assumptions made on α and β insure that this integral converges. By the Fubini theorem the exchange of order of integrations above is permissible and

(92) $$I^\alpha(I^\beta f) = \frac{H_n(\alpha+\beta)}{H_n(\alpha)H_n(\beta)} c_n(\alpha,\beta)\ I^{\alpha+\beta}f.$$

It remains to calculate $c_n(\alpha,\beta)$. For this we use the following lemma. As in §2 let $\mathcal{S}^*(\mathbb{R}^n)$ denote the set of functions in $\mathcal{S}(\mathbb{R}^n)$ which are orthogonal to all polynomials.

LEMMA 8.5. Each I^α leaves the space $\mathcal{S}^*(\mathbb{R}^n)$ invariant.

Proof. By continuity it suffices to prove this for those α for which $\alpha-n$ satisfies the assumptions of Lemma 8.2, that is, $\alpha-n, -\alpha \notin 2\mathbb{Z}^+$. But then if $f \in \mathcal{S}^*$

(93) $$(I^\alpha f)^\sim(u) = \frac{1}{H_n(\alpha)}(f * r^{\alpha-n})^\sim = \tilde{f}(u)|u|^{-\alpha},$$

since

$$(\phi * S)^\sim = \tilde{\phi}\tilde{S} \qquad \text{for } \phi \in \mathcal{S},\ S \in \mathcal{S}'.$$

But $\tilde{f}$ has all derivatives 0 at 0 and so does $\tilde{f}(u)|u|^{-\alpha}$ so the lemma is proved.

We can now finish the proof of Prop. 8.4. Taking $f_o \in \mathcal{S}^*$ we can put $f = I^\beta f_o$ in (93) and then

$$(I^{\alpha}(I^{\beta}f_0))^{\sim}(u) = (I^{\beta}f_0)^{\sim}(u)|u|^{-\alpha} = \tilde{f}_0(u)|u|^{-\alpha-\beta}$$
$$= (I^{\alpha+\beta}f_0)^{\sim}(u).$$

This shows that the scalar factor in (92) equals 1 so Prop. 8.4 is proved. In the process we have obtained the evaluation

$$\int_{\mathbf{R}^n} |v|^{\alpha-n}|e_1-v|^{\beta-n}dv = \frac{H_n(\alpha)H_n(\beta)}{H_n(\alpha+\beta)}.$$

We now prove a variation of Prop. 8.4 needed in the theory of the Radon transform.

PROPOSITION 8.6. Let $0 < k < n$. Then

$$I^{-k}(I^k f) = f \qquad \text{for all} \qquad f \in \mathcal{S}(\mathbf{R}^n).$$

Proof. By Prop. 8.4 we have

$$(94) \qquad I^{\alpha}(I^k f) = I^{\alpha+k}f \qquad \text{for} \qquad 0 < \operatorname{Re}\alpha < n-k.$$

We shall now prove, following a suggestion of R. Seeley, that the function $\phi = I^k f$ satisfies

$$(95) \qquad \sup_x |\phi(x)||x|^{n-k} < \infty .$$

For each $N > 0$ we have an estimate $|f(y)| \leq C_N(1+|y|)^{-N}$ where C_N is a constant. Thus we have

$$\left|\int_{\mathbb{R}^n} f(y)|x-y|^{k-n}dy\right| \leq C_N\int_{|y|\leq\frac{1}{2}|x|} (1+|y|)^{-N}|x-y|^{k-n}dy$$

$$+ C_N\int_{|y|\geq\frac{1}{2}|x|} (1+|y|)^{-N}|x-y|^{k-n}dy.$$

In the first integral we have $|x-y|^{k-n} \leq |\frac{1}{2}x|^{k-n}$ and in the second we use the inequality

$$(1+|y|)^{-N} \leq (1+|y|)^{-N-k+n}(1+|\tfrac{1}{2}x|)^{k-n}.$$

Taking N large enough both integrals on the right hand side will satisfy (95) so (95) is proved.

We claim now that $I^{\alpha}(\phi)(x)$, which by (94) is holomorphic for $0 < \mathrm{Re}\alpha < n-k$, extends to a holomorphic function in the half plane $\mathrm{Re}\alpha < n-k$. It suffices to prove this for $x = 0$. We decompose $\phi = \phi_1 + \phi_2$ where ϕ_1 is a smooth function identically 0 in a neighborhood $|x| < \varepsilon$ of 0, and $\phi_2 \in \mathcal{S}(\mathbb{R}^n)$. Since ϕ_1 satisfies (95) we have for $\mathrm{Re}\alpha < n-k$,

$$\left|\int\phi_1(x)|x|^{\alpha-n}dx\right| \leq C\int_{\varepsilon}^{\infty}|x|^{k-n}|x|^{\alpha-n}|x|^{n-1}d|x|$$

$$= C\int_{\varepsilon}^{\infty}|x|^{\alpha+k-n-1}d|x| < \infty$$

so $I^{\alpha}\phi_1$ is holomorphic in this half plane. On the other hand $I^{\alpha}\phi_2$ is holomorphic for $\alpha \in \mathbb{C}_n$. Now we can put $\alpha = -k$ in (94) and the proposition is proved.

BIBLIOGRAPHICAL NOTES

§§ 1-2. The inversion formulas

(i) $$f(x) = \tfrac{1}{2}(2\pi i)^{1-n} L_x^{\frac{1}{2}(n-1)} \int_{\mathbb{S}^{n-1}} J(\omega,(\omega,x))\,d\omega \qquad (n \text{ odd})$$

(ii) $$f(x) = (2\pi i)^{-n} L_x^{\frac{1}{2}(n-2)} \int_{\mathbb{S}^{n-1}} d\omega \int_{-\infty}^{\infty} \frac{dJ(\omega,p)}{p-(\omega,x)} \qquad (n \text{ even})$$

for a function $f \in C_c^\infty(\mathbb{R}^n)$ in terms of its plane integrals $J(\omega,p)$ go back to Radon [1917] and John [1934], [1955].

The support theorem, the Paley-Wiener theorem and the Schwartz theorem (Theorems 2.4, 2.6, 2.10) are from Helgason [1964A], [1965A]. The example in Remark 2.9 was also found by D.J. Newman, cf. Weiss' paper [1967], which gives another proof of the support theorem. The local result in Cor. 2.12 goes back to John [1935]; our derivation is suggested by the proof of a similar lemma in Flensted-Jensen [1977], p. 83. Another proof is in Ludwig [1966].

Corollary 2.8 was derived by Ludwig [1966] in a different way. He approaches the Schwartz - and Paley - Wiener theorems by expanding $\hat{f}(\omega,p)$ into spherical harmonics in ω. However, a crucial point in the proof is not established and seems difficult to settle in the context: the smoothness of the function $\tilde{f}(\xi)$ in (2.7), p. 57, at the point $\xi = 0$. (This is the function F in our proof of Theorem 2.4; here the smoothness of F at 0 is the main point).

Since the inversion formula (Theorem 3.1) is rather easy to prove for odd n it is natural to try to prove Theorem 2.4 for this case by showing directly that if $\phi \in \mathcal{S}_H(\mathbb{P}^n)$ then the function

$f = cL^{\frac{1}{2}(n-1)}(\check{\phi})$ belongs to $\mathcal{S}(\mathbb{R}^n)$ (in general $\check{\phi} \notin \mathcal{S}(\mathbb{R}^n)$). This approach is taken in Gelfand-Graev-Vilenkin [1962], pp. 16-17; if the analytic manipulation on page 17 could be justified one would at least get the partial result $f(x) \to 0$ for $|x| \to \infty$. Cor. 2.5 is stated in Semyanistyi [1960].

§§ 3-4. The inversion formulas (i) and (ii) above are proved in John [1955], Ch. I, by direct computations involving the solution of Poisson's equation $Lu = f$. Alternative proofs using distributions were given in Gelfand-Schilov [1959]. The dual transforms $f \to \hat{f}$, $\phi \to \check{\phi}$, the unified inversion formula and its dual

$$cf = L^{\frac{1}{2}(n-1)}((\hat{f})^{\vee}), \quad c\phi = \square^{\frac{1}{2}(n-1)}((\check{\phi})^{\wedge})$$

were given by the author in [1964A]. The proofs from Helgason [1959], p. 284 and [1965A] are based on the Darboux equation (Lemma 3.2) and therefore generalize to two-point homogeneous spaces. Formulas (17) and (55) were already given by Fuglede [1958]; according to Radon [1917], the first formula had even been observed by Herglotz. The modified inversion formula (Theorem 3.4) and Theorem 4.1 are proved in Ludwig [1966]. The latter result is attributed to Y. Reshetnyak in Gelfand-Graev-Vilenkin [1962].

§ 5. The approach to Radon transforms of distributions adopted in the text is from the author's paper [1965B]. Other methods are proposed in Gelfand-Graev-Vilenkin [1962] and in Ludwig [1966] (where formula (46) is also proved).

§ 6. The d-plane transform and Theorem 6.3 are from Helgason [1959],

p. 284. An L^2-version of Cor. 6.4 was given by Solmon [1976], p. 77. A different characterization of the image (for d=1, n=3) was given by John [1938].

Some difficulties with the d-plane transform on $L^2(\mathbb{R}^n)$ are pointed out by Smith and Solmon [1975] and Solmon [1976], p. 68. In fact, the function $f(x) = |x|^{-\frac{1}{2}n}(\log|x|)^{-1}$ $(|x| \geq 2)$, 0 otherwise, is square integrable on $\mathbb{R}^n$ but is not integrable over any plane of dimension $\geq \frac{n}{2}$.

§7. a) These applications to partial differential equations go in part back to Herglotz [1931]; our treatment is mostly based on John [1955]. Other applications of the Radon transform to partial differential equations with constant coefficients can be found in Courant-Lax [1955], Gelfand-Shapiro [1955], John [1955], Borovikov [1959], Gårding [1961], Ludwig [1966], and Lax-Phillips [1967]. Applications to general elliptic equations were given by John [1955].

While the Radon transform on $\mathbb{R}^n$ can be used to "reduce" partial differential equations to ordinary differential equations one can use a Radon type transform on a symmetric space X to "reduce" an invariant differential operator D on X to a partial differential operator with constant coefficients (cf. Helgason [1963], [1964B] and [1973]). The inversion formula gives a solutions formula for, say, the wave equation on X (1963), the Schwartz theorem analog gives a fundamental solution for D (1964B) and the Paley-Wiener theorem analog gives the surjectivity $DC^{\infty}(X) = C^{\infty}(X)$ (1973). For the case when X is a 3-dimensional hyperbolic space a related study of the wave equation is given by Lax-Phillips [1979].

b) Applications of the Radon transform in medicine were proposed by Cormack [1963] [1964] and in radio astronomy by Bracewell and Riddle [1967]. See Shepp and Kruskal [1978] for a recent survey. For the approximate reconstruction problem, including Props. 7.5, 7.6 and refinements of Theorems 7.7, 7.8 see Smith, Solmon and Wagner [1977], Solmon [1976] and Hamaker and Solmon [1978]. Theorem 7.8 is due to Halperin [1962], the proof in the text to Amemiya and Ando [1965].

§8. For a thorough treatment of distribution theory on $\mathbb{R}^n$ and on manifolds see Schwartz [1966] or Trèves [1967]. A concise, but nevertheless self-contained, treatment of the basics is given in Hörmander [1963]. A more systematic study of the potentials I^γ and the distributions r^α, x_+^α is given in Riesz [1949], Schwartz [1966], Gelfand-Schilov [1959]. However, we have not found a proof of Prop. 8.6 in the literature.

CHAPTER II

A DUALITY IN INTEGRAL GEOMETRY. GENERALIZED RADON TRANSFORMS AND ORBITAL INTEGRALS.

§1. A Duality for Homogeneous Spaces.

The inversion formulas in Theorems 3.1, 3.3, 3.4 and 3.7, Ch. I suggest the general problem of determining a function on a manifold by means of its integrals over certain submanifolds. In order to provide a natural framework for such problems we consider the Radon transform $f \to \hat{f}$ on $\mathbb{R}^n$ and its dual $\phi \to \check{\phi}$ from a group-theoretic point of view, motivated by the fact that the isometry group $M(n)$ acts transitively on both $\mathbb{R}^n$ and on the hyperplane space $\mathbb{P}^n$. Thus

(1) $$\mathbb{R}^n = M(n)/\mathbf{O}(n) \quad , \quad \mathbb{P}^n = M(n)/\,\mathbb{Z}_2 \times M(n-1),$$

where $\mathbf{O}(n)$ is the orthogonal group fixing the origin $0 \in \mathbb{R}^n$ and $\mathbb{Z}_2 \times M(n-1)$ is the subgroup of $M(n)$ leaving a certain hyperplane ξ_o through 0 stable. ($\mathbb{Z}_2$ consists of the identity and the reflection in this hyperplane.)

We observe now that a point $g_1\mathbf{O}(n)$ in the first coset space above lies on a plane $g_2(\mathbb{Z}_2 \times M(n-1))$ in the second if and only if these cosets, considered as subsets of $M(n)$, have a point in common. This leads to the following general setup.

Let G be a locally compact group, X and Ξ two left coset spaces of G,

(2) $$X = G/H_X \quad , \quad \Xi = G/H_\Xi ,$$

where H_X and H_Ξ are closed subgroups of G. It will be convenient to make the following assumptions:

(i) The groups G, H_X, H_Ξ, $H_X \cap H_\Xi$ are all unimodular. (The left invariant Haar measures are right invariant.)

(ii) If $h_X H_\Xi \subset H_\Xi H_X$ then $h_X \in H_\Xi$.

If $h_\Xi H_X \subset H_X H_\Xi$ then $h_\Xi \in H_X$.

(iii) The set $H_X H_\Xi \subset G$ is closed.

We note that (iii) is satisfied if one of the subgroups H_X, H_Ξ is compact.

All the assumptions are satisfied for the pair of coset spaces in (1). Let us for example check the first part of (ii). If $h_X H_\Xi \subset H_\Xi H_X$ we obtain by applying both sides to the origin, $h_X \cdot \xi_o \subset \xi_o$ so $h_X \in H_\Xi$.

Two elements $x \in X$, $\xi \in \Xi$ are said to be <u>incident</u> if as cosets in G they intersect. We put

$$\check{x} = \{\xi \in \Xi \; : x \text{ and } \xi \text{ incident}\},$$

$$\hat{\xi} = \{x \in X \; : x \text{ and } \xi \text{ incident}\}.$$

Using the notation $A^g = gAg^{-1}$ for $g \in G$, $A \subset G$ we have the following lemma.

LEMMA 1.1. *Let* $g \in G$, $x = gH_X$, $\xi = gH_\Xi$. *Then*

a) $\check{x}$ *is an orbit of* $(H_X)^g$ *and we have the coset space identification*

$$\check{x} = (H_X)^g/(H_X \cap H_\Xi)^g \; ;$$

b) $\hat{\xi}$ <u>is an orbit of</u> $(H_\Xi)^g$ <u>and</u>

$$\hat{\xi} = (H_\Xi)^g/(H_X \cap H_\Xi)^g.$$

<u>Proof</u>. By definition

$$\check{x} = \{\gamma H_\Xi : \gamma H_\Xi \cap gH_X \neq \emptyset\},$$

which can be written

$$\check{x} = \{ gh_X H_\Xi : h_X \in H_X \}. \tag{3}$$

This is the orbit of the point gH_Ξ in Ξ under the group gH_Xg^{-1}. The subgroup leaving the point gH_Ξ fixed is $(gH_\Xi g^{-1}) \cap (gH_X g^{-1})$. This proves a); b) follows in the same way.

Let $x_o = \{H_X\}$ and $\xi_o = \{H_\Xi\}$ denote the origins in X and Ξ, respectively. Then by (3)

$$\check{x} = g \cdot \check{x}_o \qquad \hat{\xi} = g \cdot \hat{\xi}_o,$$

where the dot $\cdot$ denotes the action of G on X and on Ξ.

LEMMA 1.2. <u>The maps</u> $x \to \check{x}$ <u>and</u> $\xi \to \hat{\xi}$ <u>are one-to-one</u>.

<u>Proof</u>. Suppose $x_1, x_2 \in X$ and $\check{x}_1 = \check{x}_2$. Let $g_1, g_2 \in G$ such that $x_1 = g_1 H_X$, $x_2 = g_2 H_X$. Then by (3) $g_1 \cdot \check{x}_o = g_2 \cdot \check{x}_o$ so writing $g = g_1^{-1} g_2$ we have $g \cdot \check{x}_o = \check{x}_o$. In particular, $g \cdot \xi_o \in \check{x}_o$ so, since $\check{x}_o$ is the orbit of $\xi_o \in \Xi$ under H_X, we have $g \cdot \xi_o = h_X \cdot \xi_o$ for some $h_X \in H_X$, whence $h_X^{-1} g = h_\Xi \in H_\Xi$. It follows that $h_\Xi \cdot \check{x}_o = \check{x}_o$ so $h_\Xi H_X \cdot \xi_o = H_X \cdot \xi_o$, that is $h_\Xi H_X \subset H_X H_\Xi$. By assumption (ii) $h_\Xi \in H_X$ which gives $x_1 = x_2$. This proves the lemma.

In view of this lemma, X and Ξ are homogeneous spaces of the same group G such that each point in Ξ can be viewed as a subset of X and each point of X can be viewed as a subset of Ξ. We say X and Ξ are *homogeneous spaces in duality*. The terminology is suggested by the familiar duality between $\mathbb{R}^n$ and $\mathbb{P}^n$ in Projective Geometry.

The maps $x \to \check{x}$ and $\xi \to \hat{\xi}$ are also conveniently described by means of the following double fibration,

$$
\begin{array}{ccc}
 & G/(H_X \cap H_\Xi) & \\
p \swarrow & & \searrow \pi \\
X = G/H_X & & \Xi = G/H_\Xi ,
\end{array}
\tag{4}
$$

where the maps p and π are given by $p(gH_X \cap H_\Xi) = gH_X$, $\pi(gH_X \cap H_\Xi) = gH_\Xi$. Then by (3) we have

$$\check{x} = \pi(p^{-1}(x)) , \qquad \hat{\xi} = p(\pi^{-1}(\xi)). \tag{5}$$

LEMMA 1.3. *Each* $\check{x} \subset \Xi$ *is closed and each* $\hat{\xi} \subset X$ *is closed.*

Proof. If $p_\Xi : G \to G/H_\Xi$ is the natural mapping we have

$$(p_\Xi)^{-1}(\Xi - (\check{x}_o)) = \{g : gH_\Xi \notin H_X \cdot H_\Xi\} = G - H_X H_\Xi .$$

In particular, $p_\Xi(G - H_X H_\Xi) = \Xi - \check{x}_o$, so using (ii) and the fact that p_Ξ is an open mapping we deduce that $\check{x}_o$ is closed. By translation each $\check{x}$ is closed and similarly each $\hat{\xi}$ is closed.

Examples. (i) *Points outside hyperplanes.* We saw before that if in the coset space representation (1) $\mathbf{O}(n)$ is viewed as the isotropy group of 0 and $\mathbb{Z}_2 M(n-1)$ is viewed as the isotropy group of a hyperplane *through* 0 then the abstract incidence notion is equivalent to the naive one: $x \in \mathbb{R}^n$ is incident to $\xi \in \mathbb{P}^n$ if and only if $x \in \xi$.

On the other hand we can also view $\mathbb{Z}_2 M(n-1)$ as the isotropy group of a hyperplane ξ_δ at a distance $\delta > 0$ from 0. (This amounts to a different imbedding of the group $\mathbb{Z}_2 M(n-1)$ into $M(n)$). Then we have the following generalization.

PROPOSITION 1.4. *The point* $x \in \mathbb{R}^n$ *and the hyperplane* $\xi \in \mathbb{P}^n$ *are incident if and only if distance* $(x, \xi) = \delta$.

Proof. Let $x = gH_X$, $\xi = \gamma H_\Xi$ where $H_X = \mathbf{O}(n)$, $H_\Xi = \mathbb{Z}_2 M(n-1)$. Then if $gH_X \cap \gamma H_\Xi \neq \emptyset$, we have $gh_X = \gamma h_\Xi$ for some $h_X \in H_X$, $h_\Xi \in H_\Xi$. Now the orbit $H_\Xi \cdot 0$ consists of the two planes ξ'_δ and ξ''_δ parallel to ξ_δ at a distance δ from ξ_δ. The relation $g \cdot 0 = \gamma h_\Xi \cdot 0 \in \gamma \cdot (\xi'_\delta \cup \xi''_\delta)$ together with the fact that g and γ are isometries shows that x has distance δ from $\gamma \cdot \xi_\delta = \xi$.

On the other hand if distance $(x, \xi) = \delta$ we have $g \cdot 0 \in \gamma \cdot (\xi'_\delta \cup \xi''_\delta) = \gamma H_\Xi \cdot 0$ which means $gH_X \cap \gamma H_\Xi \neq \emptyset$.

(ii) *Unit spheres.* Let σ_o be a sphere in $\mathbb{R}^n$ of radius one passing through the origin. Denoting by Σ the set of all *unit* spheres in $\mathbb{R}^n$ we have the dual homogeneous spaces

(6) $$\mathbb{R}^n = M(n)/\mathbf{O}(n) \quad ; \quad \Sigma = M(n)/\mathbf{O}^*(n)$$

where $\mathbf{O}^*(n)$ is the set of rotations around the center of σ_o. Here a point $x = g\,\mathbf{O}(n)$ is incident to $\sigma = \gamma\,\mathbf{O}^*(n)$ if and only if $x \in \sigma$.

(iii) *Complex flag manifolds.* Consider the complex n-space $\mathbb{C}^n$ and a fixed set of integers $0 < d_1 < \ldots < d_r < n$. A flag in $\mathbb{C}^n$ of type $(d_1, \ldots, d_r)$ is a sequence of subspaces $L_1 \subset L_2 \subset \ldots \subset L_r \subset \mathbb{C}^n$ where $\dim L_i = d_i$ $(1 \leq i \leq r)$. The set $F_{d_1 \ldots d_r}$ of all flags in $\mathbb{C}^n$ of type $(d_1, \ldots, d_r)$ is a complex manifold in a natural way: In fact the general linear group $GL(n,\mathbb{C})$ acts transitively on it with a complex isotropy group.

Consider in particular the manifold F_{12} of flags of type (1,2) in $\mathbb{C}^4$, the complex projective space $F_1 = P_3(\mathbb{C})$ (of complex lines in $\mathbb{C}^4$) and the Grassmann manifold $F_2 = G_{2,4}(\mathbb{C})$ of complex two-dimensional subspaces of $\mathbb{C}^4$. The group $\mathbb{U}(4)$ acts transitively on these manifolds F_{12}, F_1 and F_2 and our double fibration (4) becomes

$$F_{12} = \mathbb{U}(4)/(\mathbb{U}(1)\times\mathbb{U}(1)\times\mathbb{U}(2))$$

$$F_1 = \mathbb{U}(4)/(\mathbb{U}(1)\times\mathbb{U}(3)) \qquad\qquad F_2 = \mathbb{U}(4)/(\mathbb{U}(2)\times\mathbb{U}(2))$$

and our maps $x \to \check{x}$, $\xi \to \hat{\xi}$ mean the following:

A. The map $x \to \check{x}$ associates to a complex line the set of 2-dimensional complex subspaces $\subset \mathbb{C}^4$ containing it.

B. The map $\xi \to \hat{\xi}$ associates to a 2-dimensional complex subspace of $\mathbb{C}^4$ the set of complex lines contained in it.

These maps A,B have been utilized by Penrose; they are called Penrose correspondences in the paper Wells [1979] to which refer for an account of their applications to certain differential equations in mathematical physics.

§2. The Radon Transform for the Double Fibration (4).

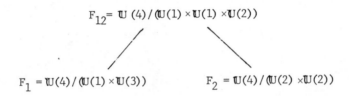

In accordance with the unimodularity assumption (i) we fix invariant measures dg, dh_X, dh_Ξ and dh on the groups G, H_X, H_Ξ and $H = H_X \cap H_\Xi$, respectively. By the existence theorem for invariant measures on homogeneous spaces there exists a unique H_X- invariant measure $d\mu = d(h_X)_H$ on $\check{x}_0 = H_X/H$ satisfying

$$\int_{H_X} f(h_X)\, dh_X = \int_{\check{x}_0} \left(\int_H f(h_X h)\, dh \right) d\mu(h_X H) \tag{7}$$

for all $f \in C_c(H_X)$. Similarly, there is defined an H_Ξ - invariant measure dm on $\hat{\xi}_0 = H_\Xi / H$. We shall now see that by translations we obtain consistently defined invariant measures on each $\check{x}$.

LEMMA 2.1. *There exists a nonzero measure on each* $\check{x}$, *coinciding with* $d\mu$ *on* $\check{x}_0$ *such that whenever* $g \cdot \check{x}_1 = \check{x}_2$ *the measures correspond under* g.

Similar statement holds for $\hat{\xi}$.

Proof. If $\check{x} = g \cdot \check{x}_0$ we transfer the measure $d\mu$ on $\check{x}_0$ over to $\check{x}$ by means of the homeomorphism $\xi \to g \cdot \xi$. This gives an $(H_X)^g$-invariant measure on $\check{x}$ but since such measures are only determined up to a constant factor we must still prove independence of the choice of g. But if $g' \cdot \check{x}_0 = g \cdot \check{x}_0$ we have $(g \cdot x_0)^\vee = (g' \cdot x_0)^\vee$. Thus by Lemma 1.2, $g \cdot x_0 = g' \cdot x_0$ whence $g \in g' H_X$. Since $d\mu$ on $\check{x}_0$ is H_X-invariant the lemma follows.

The measures on $\check{x}$ and $\hat{\xi}$ defined by Lemma 2.1 will be denoted by $d\mu$ and dm, respectively. We also denote by dg_{H_X} and dg_{H_Ξ}, respectively, the G-invariant measures on X and Ξ which are normalized by the relations

$$\int_G F(g)\, dg = \int_X \left(\int_{H_X} F(gh_X)\, dh_X \right) dg_{H_X} , \tag{8}$$

$$\int_G \Phi(g)\,dg = \int_\Xi \left(\int_{H_\Xi} \Phi(gh_\Xi)\,dh_\Xi\right) dg_{H_\Xi}. \tag{9}$$

We shall also put $dx = dg_{H_X}$, $d\xi = dg_{H_\Xi}$ for simplicity. We now define the Radon transform $f \to \hat{f}$ and its dual $\phi \to \check{\phi}$ by the formulas

$$\hat{f}(\xi) = \int_{\hat{\xi}} f(x)\,dm(x), \quad \check{\phi}(x) = \int_{\check{x}} \phi(\xi)\,d\mu(\xi) \tag{10}$$

for $f \in C_c(X)$, $\phi \in C_c(\Xi)$. Since $\hat{\xi}$ is closed the restriction $f|\hat{\xi}$ belongs to $C_c(\hat{\xi})$. Thus the integrals (10) are well defined. Formula (10) can also be written in group-theoretic terms:

$$\hat{f}(gH_\Xi) = \int_{H_\Xi/H} f(gh_\Xi H_X)\,d(h_\Xi)_H, \quad \check{\phi}(gH_X) = \int_{H_X/H} \phi(gh_X H_\Xi)\,d(h_X)_H. \tag{11}$$

PROPOSITION 2.2. Let $f \in C_c(X)$, $\phi \in C_c(\Xi)$. Then $\hat{f}$ and $\check{\phi}$ are continuous and

$$\int_X f(x)\,\check{\phi}(x)\,dx = \int_\Xi \hat{f}(\xi)\,\phi(\xi)\,d\xi.$$

Proof. The continuity of $\hat{f}$ (and of $\check{\phi}$) is immediate from (11). Next we consider the double fibration (4) where $p(gH) = gH_X$, $\pi(gH) = gH_\Xi$. We fix a G-invariant measure dg_H on G/H such that for all $F \in C_c(G)$

$$\int_G F(g)\,dg = \int_{G/H} \left(\int_H F(gh)\,dh\right) dg_H. \tag{12}$$

By the "chain rule" for invariant integration we have

(13) $$\int_{G/H} Q(gH)\,dg_H = c\int_{G/H_X} dg_{H_X} \int_{H_X/H} Q(gh_X H)\,d(h_X)_H ,$$

where $Q \in C_c(G/H)$ is arbitrary and c is a constant independent of Q. Combining (12) and (13) we obtain

(14) $$\int_G F(g)\,dg = c\int_{G/H_X} dg_{H_X} \int_{H_X/H} d(h_X)_H \int_H F(gh_X h)\,dh ,$$

which by (7) equals

$$c\int_{G/H_X} dg_{H_X} \int_{H_X} F(gh_X)\,dh_X .$$

Comparing with (8) we obtain $c=1$.

We consider now the function

$$P = (f\circ p)(\phi\circ\pi)$$

on G/H. We integrate it over G/H in two ways corresponding to the two fibrations in (6). This amounts to using (13) and its analog for H_Ξ. The result is

(15) $$\int_{G/H} P(gH)\,dg_H = \int_{G/H_X} dg_{H_X} \int_{H_X/H} P(gh_X H)\,d(h_X)_H ,$$

(16) $$\int_{G/H} P(gH)\,dg_H = \int_{G/H_\Xi} dg_{H_\Xi} \int_{H_\Xi/H} P(gh_\Xi H)\,d(h_\Xi)_H$$

and as in the Fubini theorem absolute convergence of one side implies that of the other. But

$$P(gh_X H) = f(gH_X)\,\phi(gh_X H_\Xi)$$

and

$$\int_{H_X/H} \phi(gh_X H_\Xi)\,d(h_X)_H = \check{\phi}(gH_X)$$

so the right hand side of (15) reduces to

$$\int_X f(x)\,\check{\phi}(x)\,dx\ .$$

Treating (16) similarly we obtain the proposition.

The result shows how to define the Radon transform and its dual for measures and, in case G is a Lie group, for distributions.

Definition. Let s be a measure on X of compact support. Its Radon transform is the functional $\hat{s}$ on $C_c(\Xi)$ defined by

$$\hat{s}(\phi) = s(\check{\phi}). \tag{17}$$

Similarly $\check{\sigma}$ is defined by

$$\check{\sigma}(f) = \sigma(\hat{f}) \qquad f \in C_c(X) \tag{18}$$

if σ is a compactly supported measure on Ξ.

LEMMA 2.3. (i) *If s is a compactly supported measure on X, $\hat{s}$ is a measure on Ξ.*

(ii) *If s is a bounded measure on X and if $\check{x}_o$ has finite measure then $\hat{s}$ as defined by (17) is a bounded measure.*

Proof. (i) The measure s can be written as a difference $s = s^+ - s^-$ of two positive measures, each of compact support. Then $\hat{s} = \hat{s^+} - \hat{s^-}$ is a difference of two positive functionals on $C_c(\Xi)$.

Since a positive functional is necessarily a measure, $\hat{s}$ is a measure.

(ii) We have

$$\sup_{x} |\check{\phi}(x)| \leq \sup_{\xi} |\phi(\xi)| \; \mu(\check{x}_0)$$

so for a constant K,

$$|\hat{s}(\phi)| = |s(\check{\phi})| \leq K \sup|\check{\phi}| \leq K\mu(\check{x}_0) \sup|\phi| ,$$

and the boundedness of $\hat{s}$ follows.

If G is a Lie group then (17), (18) with $f \in \mathcal{D}(X)$, $\phi \in \mathcal{D}(\Xi)$ serve to define the Radon transform $s \to \hat{s}$ and the dual $\sigma \to \check{\sigma}$ for distributions s and σ of compact support. We consider the spaces $\mathcal{D}(X)$ and $\mathcal{E}(X)$ $(= C^{\infty}(X))$ with their customary topologies (Schwartz [1966], Ch. III). The duals $\mathcal{D}'(X)$ and $\mathcal{E}'(X)$ then consist of the distributions on X and the distributions on X of compact support, respectively.

PROPOSITION 2.4. _The mappings_

$$f \in \mathcal{D}(X) \to \hat{f} \in \mathcal{E}(\Xi)$$

$$\phi \in \mathcal{D}(\Xi) \to \check{\phi} \in \mathcal{E}(X)$$

are continuous. In particular,

$$s \in \mathcal{E}'(X) \Rightarrow \hat{s} \in \mathcal{D}'(\Xi)$$

$$\sigma \in \mathcal{E}'(\Xi) \Rightarrow \check{\sigma} \in \mathcal{D}'(X) .$$

Proof. We have

$$(19) \qquad \hat{f}(g \cdot \xi_0) = \int_{\hat{\xi}_0} f(g \cdot x)\, dm(x).$$

Let g run through a local cross section through e in G over a neighborhood of ξ_0 in Ξ. If $(t_1, \ldots, t_n)$ are coordinates of g and $(x_1, \ldots, x_m)$ the coordinates of $x \in \hat{\xi}_0$ then (19) can be written in the form

$$\hat{F}(t_1, \ldots, t_n) = \int F(t_1, \ldots, t_n \,;\, x_1, \ldots, x_m)\, dx_1 \ldots dx_m.$$

Now it is clear that $\hat{f} \in \mathcal{E}(\Xi)$ and that $f \to \hat{f}$ is continuous, proving the proposition.

The results for the Radon transform on $\mathbb{R}^n$ in Chapter I now suggest the following general problems:

A. Relate function spaces on X and on Ξ by means of the integral transforms $f \to \hat{f}, \phi \to \check{\phi}$. In particular, relate the supports of f and of $\hat{f}$.

B. Relate directly the functions f and $(\hat{f})^{\vee}$ on X, the functions ϕ and $(\check{\phi})^{\wedge}$ on Ξ.

C. In case G is a Lie group, let $\mathbb{D}(X)$ and $\mathbb{D}(\Xi)$ denote the algebras of G-invariant differential operators on X and Ξ, respectively.

Does there exist a map $D \to \hat{D}$ of $\mathbb{D}(X)$ into $\mathbb{D}(\Xi)$ and a map $E \to \check{E}$ of $\mathbb{D}(\Xi)$ into $\mathbb{D}(X)$ such that

$$(Df)^{\wedge} = \hat{D}\hat{f}\,, \quad (E\phi)^{\vee} = \check{E}\check{\phi} \qquad ?$$

Theorem 2.4, 2.6, 2.10, 3.1, 3.3, Lemma 2.1 and Cor. 2.5 in Ch.I give answers to these questions for $X = \mathbb{R}^n$, $\Xi = \mathbb{P}^n$. But while the problems can be posed for the general double fibration (4) one can not expect complete solutions in this generality. For example, if $f \to \hat{f}$ is the Radon transform on S^2 defined by integration over geodesics the function $(\hat{f})^\vee$ will, in contrast to f, necessarily be a symmetric function. Here however assumption (ii) does not hold.

In the next section we consider Problems A and B in detail for the case when X is a two-point homogeneous space, or equivalently, an isotropic Riemannian manifold.

Examples. (i) Let G denote the group $SL(2,\mathbb{R})$ of 2×2 matrices of determinant one and Γ the modular group $SL(2,\mathbb{Z})$. Let N denote the unipotent group $\begin{pmatrix} 1 & n \\ 0 & 1 \end{pmatrix}$ where $n \in \mathbb{R}$ and consider the homogeneous spaces

$$(20) \qquad X = G/N\,, \qquad \Xi = G/\Gamma\,.$$

Under the usual action of G on $\mathbb{R}^2$ N is the isotropy subgroup of $(1,0)$ so X can be identified with $\mathbb{R}^2-(0)$, whereas Ξ is of course three-dimensional.

In number theory one is interested in decomposing the space $L^2(G/\Gamma)$ into G-invariant irreducible subspaces. We now give a rough description of this by means of the transforms $f \to \hat{f}$ and $\phi \to \check{\phi}$.

As customary we put $\Gamma_\infty = \Gamma \cap N$; our transforms (10) then take the form

$$\hat{f}(g\Gamma) = \sum_{\Gamma/\Gamma_\infty} f(g\gamma N), \qquad \check{\phi}(gN) = \int_{N/\Gamma_\infty} \phi(gn\Gamma)\, dn_{\Gamma_\infty}\,.$$

Since N/Γ_∞ is the circle group, $\check{\phi}(gN)$ is just the constant term in the Fourier expansion of the function $n\Gamma_\infty \to \phi(gn\Gamma)$. The null space $L^2_d(G/\Gamma)$ in $L^2(G/\Gamma)$ of the operator $\phi \to \check{\phi}$ is called the space of cusp forms. According to Prop. 2.2 they constitute the orthogonal complement of the image $C_c(X)^{\wedge}$.

We have now the G-invariant decomposition

(21) $$L^2(G/\Gamma) = L^2_c(G/\Gamma) \oplus L^2_d(G/\Gamma) ,$$

where (— denoting closure)

(22) $$L^2_c(G/\Gamma) = (C_c(X)^{\wedge})^{-}$$

and as mentioned above,

(23) $$L^2_d(G/\Gamma) = (C_c(X)^{\wedge})^{\perp} .$$

It is known (cf. Selberg [1962], Godement [1966]) that the representation of G on $L^2_c(G/\Gamma)$ is the continuous direct sum of the irreducible representations of G from the principal series whereas the representation of G on $L^2_d(G/\Gamma)$ is the discrete direct sum of irreducible representations each occuring with finite multiplicity.

(ii) The determination of a function in $\mathbb{R}^n$ in terms of its integrals over unit spheres (John [1955]) can be regarded as a solution to the first half of Problem B for the double fibration (6).

§ 3. Orbital Integrals.

As before let $X = G/H_X$ be a homogeneous space with origin $o = \{H_X\}$. Given $x_o \in X$ let G_{x_o} denote the subgroup of G leaving x_o fixed, i.e. the isotropy subgroup of G at x_o.

Definition. A generalized sphere is an orbit $G_{x_o} \cdot x$ in X of some point $x \in X$ under the isotropy subgroup at some point $x_0 \in X$.

Examples. (i) If $X = \mathbb{R}^n$, $G = M(n)$ then the generalized spheres are just the spheres.

(ii) Let X be a locally compact group L and G the product group $L \times L$ acting on L on the right and left, the element $(l_1, l_2) \in L \times L$ inducing the action $l \to l_1 l l_2^{-1}$ on L. Let ΔL denote the diagonal in $L \times L$. If $l_o \in L$ then the isotropy subgroup of l_o is given by

$$(L \times L)_{l_o} = (l_o, e) \Delta L (l_o^{-1}, e)$$

and the orbit of l under it by

$$(L \times L)_{l_o} \cdot l = l_o (l_o^{-1} l)^L ,$$

that is the left translate by l_o of the conjugacy class of the element $l_o^{-1} l$. Thus the generalized spheres in the group L are the left (or right) translates of its conjugacy classes.

Coming back to the general case $X = G/H_X = G/G_o$ we assume that G_o, and therefore each G_{x_o}, is unimodular. But $G_{x_o} \cdot x = G_{x_o} / (G_{x_o})_x$ so $(G_{x_o})_x$ unimodular implies the orbit $G_{x_o} \cdot x$ has an invariant measure determined up to a constant factor. We can now consider the following general problem (following A, B, C above).

D. Determine a function f on X in terms of its integrals over generalized spheres.

Remarks. In this problem it is of course significant how the invariant measures on the various orbits are normalized.

a) If G_o is compact the problem above is rather trivial because each orbit $G_{x_o} \cdot x$ has finite invariant measure so $f(x_o)$ is given as the limit as $x \to x_o$ of the average of f over $G_{x_o} \cdot x$.

b) Suppose that for each $x_o \in X$ there is a G_{x_o}-invariant open set $C_{x_o} \subset X$ containing x_o in its closure such that for each $x \in C_{x_o}$

the isotropy group $(G_{x_o})_x$ is compact. The invariant measure on the orbit $G_{x_o}\cdot x$ $(x_o \in X,\ x \in C_{x_o})$ can then be consistently normalized as follows: Fix a Haar measure dg_o on G_o. If $x_o = g\cdot o$ we have $G_{x_o} = gG_og^{-1}$ and can carry dg_o over to a measure dg_{x_o} on G_{x_o} by means of the conjugation $z \to gzg^{-1}$ $(z \in G_o)$. Since dg_o is bi-invariant, dg_{x_o} is independent of the choice of g satisfying $x_o = g\cdot o$, and is bi-invariant. Since $(G_{x_o})_x$ is compact it has a unique Haar measure $dg_{x_o,x}$ with total measure 1 and now dg_{x_o} and $dg_{x_o,x}$ determine canonically an invariant measure μ on the orbit $G_{x_o}\cdot x = G_{x_o}/(G_{x_o})_x$. We can therefore state Problem D in a more specific form.

D') <u>Express</u> $f(x_o)$ <u>in terms of the integrals</u>

$$\int_{G_{x_o}\cdot x} f(p)\,d\mu(p) \qquad\qquad x \in C_{x_o}.$$

For the case when X is an <u>isotropic Lorentz manifold</u> the assumptions above are satisfied (with C_{x_o} consisting of the "timelike" rays from x_o) and we shall obtain in Ch. IV an explicit solution to Problem D' (Theorem 4.1, Ch. IV).

c) If in Example (ii) above L is a semisimple Lie group Problem D is a basic step in proving the Plancherel formula for the Fourier transform on L (Gelfand - Graev [1955], Harish - Chandra [1957]).

BIBLIOGRAPHICAL NOTES

The notion of <u>incidence</u> in a pair of homogeneous spaces goes back to Chern [1942]. The duality between

(i) $$G/H_X \quad \text{and} \quad G/H_\Xi,$$

the <u>correspondences</u>

$$\text{(ii)} \qquad x \to \check{x}\,, \quad \xi \to \hat{\xi}$$

and the <u>Radon transforms for a double fibration</u> of a homogeneous space were introduced in the author's paper [1965 B] from which most of this chapter comes. A further generalization, replacing the homogeneity assumption by postulates about the compatibility of the entering measures, was given by Gelfand, Graev and Shapiro [1969].

For the case $G = \mathbb{U}(4)$, $H_X = \mathbb{U}(1) \times \mathbb{U}(3)$ $H_\Xi = \mathbb{U}(2) \times \mathbb{U}(2)$ the maps (ii) become Penrose correspondences (Penrose [1967]) described in more detail in Example (iii) in §1.

The example where X is the set of p-planes in $\mathbb{R}^n$, Ξ the set of q-planes and $n=p+q+1$ is worked out in detail in Helgason [1965A] with solutions to Problems A, B, and C.

CHAPTER III

THE RADON TRANSFORM ON TWO-POINT HOMOGENEOUS SPACES

Let X be a complete Riemannian manifold, x a point in X and X_x the tangent space to X at x. Let Exp_x denote the mapping of X_x into X given by $\mathrm{Exp}_x(u) = \gamma_u(1)$ where $t \to \gamma_u(t)$ is the geodesic in X through x with tangent vector u at $x = \gamma_u(0)$.

A connected submanifold S of a Riemannian manifold X is said to be <u>totally geodesic</u> if each geodesic in X which is tangential to S at a point lies entirely in S.

The totally geodesic submanifolds of $\mathbb{R}^n$ are the planes in $\mathbb{R}^n$. Therefore, in generalizing the Radon transform to Riemannian manifolds, it is natural to consider integration over totally geodesic submanifolds. In order to have enough totally geodesic submanifolds at our disposal we consider in this section Riemannian manifolds X which are <u>two-point homogeneous</u> in the sense that for any two point pairs $p,q \in X$ $p',q' \in X$, satisfying $d(p,q) = d(p',q')$, (where d = distance), there exists an isometry g of X such that $g \cdot p = p'$, $g \cdot q = q'$. We start with the subclass of Riemannian manifolds with the richest supply of totally geodesic submanifolds, namely the spaces of constant curvature.

§1. <u>Spaces of Constant Curvature</u>.

Let X be a simply connected complete Riemannian manifold of dimension $n \geq 2$ and constant sectional curvature.

LEMMA 1.1. <u>Let</u> $x \in X$, V <u>a subspace of the tangent space</u> X_x.

<u>Then</u> $\mathrm{Exp}_x(V)$ <u>is a totally geodesic submanifold of</u> X.

<u>Proof</u>. For this we choose a specific imbedding of X into $\mathbb{R}^{n+1}$, and assume for simplicity the curvature is $\varepsilon\,(=\pm 1)$. Consider the quadratic form

$$B_\varepsilon(x) = x_1^2 + \ldots + x_n^2 + \varepsilon\, x_{n+1}^2$$

and the quadric Q_ε given by $B_\varepsilon(x) = \varepsilon$. The orthogoanl group $\mathbb{O}(B_\varepsilon)$ acts transitively on Q_ε. The form B_ε is positive definite on the tangent space $\mathbb{R}^n \times (0)$ to Q_ε at $x^o = (0, \ldots, 0,1)$; by the transitivity B_ε induces a positive definite quadratic form at each point of Q_ε, turning Q_ε into a Riemannian manifold, on which $\mathbb{O}(B_\varepsilon)$ acts as a transitive group of isometries. The isotropy subgroup at the point x^o is isomorphic to $\mathbb{O}(n)$ and it acts transitively on the set of two-dimensional subspaces of the tangent space $(Q_\varepsilon)_{x^o}$. It follows that all sectional curvatures at x^o are the same, namely ε, so by the homogeneity, Q_ε has constant curvature ε. In order to work with connected manifolds, we replace Q_{-1} by its intersection Q^+_{-1} with the half space $x_{n+1} > 0$. Then Q_{+1} and Q^+_{-1} are simply connected complete Riemannian manifolds of constant curvature. Since such manifolds are uniquely determined by the dimension and the curvature it follows that we can identify X with Q_{+1} or Q^+_{-1}.

The geodesic in X through x^o with tangent vector $(1,0,\ldots,0)$ will be left point-wise fixed by the isometry $(x_1,x_2,\ldots,x_n,x_{n+1}) \to (x_1,-x_2,\ldots,-x_n,x_{n+1})$. This geodesic is therefore the intersection of X with the two-plane $x_2 = \ldots = x_n = 0$ in $\mathbb{R}^{n+1}$. By the transitivity of $\mathbb{O}(n)$ all geodesics in X through x^o are intersections of X with two-planes through 0. By the transitivity of $\mathbf{O}(Q_\varepsilon)$ it then follows that the geodesics in X are precisely the nonempty intersections of X with two-planes through the origin.

Now if $V \subset X_{x^o}$ is a subspace, $\mathrm{Exp}_{x^o}(V)$ is by the above the intersection of X with the subspace of $\mathbb{R}^{n+1}$ spanned by V and x^o. Thus $\mathrm{Exp}_{x^o}(V)$ is a quadric in $V + \mathbb{R}x^o$ and its Riemannian structure induced by X is the same as induced by the restriction $B_\varepsilon|(V+\mathbb{R}x^o)$. Thus, by the above, the geodesics in $\mathrm{Exp}_{x^o}(V)$ are obtained by intersecting it with two-planes in $V + \mathbb{R}x^o$ through 0. Consequently, the geodesics in $\mathrm{Exp}_{x^o}(V)$ are geodesics in X so $\mathrm{Exp}_{x^o}(V)$ is totally geodesic submanifold of X. By the homogeneity of X this holds with x^o replaced by an arbitrary point $x \in X$. The lemma is proved.

In accordance with the viewpoint of Ch. II we consider X as a homogeneous space of the identity component G of the group $\mathbf{O}(Q_\varepsilon)$. Let H_X denote the isotropy subgroup of G at the point $x^o = (0, \ldots, 0, 1)$. Then H_X can be identified with the special orthogonal group $\mathbf{SO}(n)$. Let k be a fixed integer, $1 \leqslant k \leqslant n-1$; let $\xi_o \subset X$ be a fixed totally geodesic submanifold of dimension k passing through x^o and let H_Ξ be the subgroup of G leaving ξ_o invariant. We have then

$$(1) \qquad X = G/H_X\,, \qquad \Xi = G/H_\Xi\,,$$

Ξ denoting the set of totally geodesic k-dimensional submanifolds of X. Since $x^o \in \xi_o$ it is clear that the abstract incidence notion boils down to the naive one, in other words: The cosets $x = gH_X$ $\xi = \gamma H_\Xi$ have a point in common if and only if $x \in \xi$.

A. <u>The Hyperbolic Space</u>. We take first the case of negative curvature, that is $\varepsilon = -1$. The transform $f \to \hat{f}$ is now given by

(2) $$\hat{f}(\xi) = \int_{\xi} f(x)\,dm(x)$$

ξ being any k-dimensional totally geodesic submanifold of X $(1 \leqslant k \leqslant n-1)$ with the induced Riemannian structure and dm the corresponding measure. From our description of the geodesics in X it is clear that any two points in X can be joined by a unique geodesic. Let d be a distance function on X, and for simplicity we write o for the origin x^o in X. Consider now geodesic polar-coordinates for X at o; this is a mapping

$$\mathrm{Exp}_o Y \longrightarrow (r, \theta_1, \ldots, \theta_{n-1}),$$

where Y runs through the tangent space X_o, $r = |Y|$ (the norm given by the Riemannian structure) and $(\theta_1, \ldots, \theta_{n-1})$ are coordinates of the unit vector $Y/|Y|$. Then the Riemannian structure of X is given by

(3) $$ds^2 = dr^2 + (\sinh r)^2\, d\sigma^2,$$

where $d\sigma^2$ is the Riemannian structure

$$\sum_{i,j=1}^{n-1} g_{ij}(\theta_1, \ldots, \theta_{n-1})\, d\theta_i\, d\theta_j$$

on the unit sphere in X_o. The surface area $A(r)$ and volume $V(r) = \int_0^r A(t)\,dt$ of a sphere in X of radius r are thus given by

(4) $$A(r) = \Omega_n (\sinh r)^{n-1}, \quad V(r) = \Omega_n \int_0^r \sinh^{n-1} t\, dt$$

so $V(r)$ increases like $e^{(n-1)r}$. This explains the growth condition in the next result where $d(o,\xi)$ denotes the distance of o to the manifold ξ.

THEOREM 1.2. (*The support theorem.*) *Suppose* $f \in C(X)$ *satisfies*

(i) *For each integer* $m > 0$, $f(x)e^{md(o,x)}$ *is bounded.*

(ii) *There exists a number* $R > 0$ *such that*

$$\hat{f}(\xi) = 0 \qquad \text{for } d(o,\xi) > R.$$

Then

$$f(x) = 0 \qquad \text{for } d(o,x) > R.$$

Taking $R \to 0$ we obtain the following conseqwuence.

COROLLARY 1.3. *The Radon transform* $f \to \hat{f}$ *is one-to-one on the space of continuous functions on* X *satisfying condition* (i) *of "exponential decrease".*

Proof of Theorem 1.2. Using smoothing of the form

$$\int_G \phi(g) f(g^{-1} \cdot x)\, dg$$

($\phi \in C_c^\infty(G)$, dg Haar measure on G) we can (as in Theorem 2.6, Ch.I) assume that $f \in C^\infty(X)$.

We first consider the case when f in (2) is a radial function. Let P denote the point in ξ at the minimum distance $p = d(o,\xi)$ from o, let $Q \in \xi$ be arbitrary and let

$$q = d(o,Q), \qquad r = d(P,Q).$$

Since ξ is totally geodesic $d(P,Q)$ is also the distance between P and Q in ξ. Consider now the totally geodesic plane π through the geodesics oP and oQ as given by Lemma 1.1. Since a totally geodesic submanifold contains the geodesic joining any two of its points, π contains the geodesic PQ. The angle OPQ being 90° (see e.g. Helgason [1978], p.77) we conclude by hyperbolic trigonometry, (see e.g. Coxeter [1957])

$$\cosh q = \cosh p \ \cosh r. \tag{5}$$

Since f is radial it follows from (5) that the restriction $f|\xi$ is constant on spheres in ξ with center P. Since these have area $\Omega_k (\sinh r)^{k-1}$ formula (2) takes the form

$$\hat{f}(\xi) = \Omega_k \int_0^{\infty} f(Q)(\sinh r)^{k-1} dr. \tag{6}$$

Since f is a radial function it is invariant under the subgroup $H_X \subset G$ which fixes o. But H_X is not only transitive on each sphere $S^r(o)$ with center o, it is for each fixed k transitive on the set of k-dimensional totally geodesic submanifolds which are tangent to $S^r(o)$. Consequently, $\hat{f}(\xi)$ depends only on the distance $d(o,\xi)$. Thus we can write

$$f(Q) = F(\cosh q), \quad \hat{f}(\xi) = \hat{F}(\cosh p)$$

for certain 1-variable functions F and $\hat{F}$, so by (5) we obtain

$$\hat{F}(\cosh p) = \Omega_k \int_0^{\infty} F(\cosh p \cosh r)(\sinh r)^{k-1} dr. \tag{7}$$

Writing here $t = \cosh p$, $s = \cosh r$ this reduces to

$$\hat{F}(t) = \Omega_k \int_1^{\infty} F(ts)(s^2-1)^{\frac{1}{2}k-1} ds . \tag{8}$$

Here we substitute $u = (ts)^{-1}$ and then put $v = t^{-1}$. Then (8) becomes

$$v^{-1}\hat{F}(v^{-1}) = \Omega_k \int_0^v \{F(u^{-1})u^{-k}\}(v^2-u^2)^{\frac{1}{2}k-1}\,du.$$

This integral equation is of the form (19), Ch. I so we get the following analog of (20), Ch. I:

$$F(u^{-1})u^{-k} = cu\left(\frac{d}{d(u^2)}\right)^k \int_0^u (u^2-v^2)^{\frac{1}{2}k-1}\,\hat{F}(v^{-1})\,dv. \tag{9}$$

Now by assumption (ii) $\hat{F}(\cosh p) = 0$ if $p > R$. Thus $\hat{F}(v^{-1}) = 0$ if $0 < v < (\cosh R)^{-1}$. From (9) we can then conclude $F(u^{-1}) = 0$ if $u < (\cosh R)^{-1}$ which means $f(x) = 0$ for $d(o,x) > R$. This proves the theorem for f radial.

Next we consider an arbitrary $f \in C^\infty(X)$ satisfying (i), (ii). Fix $x \in X$ and if dk is the normalized Haar measure on H_x consider the integral

$$F_x(y) = \int_{H_x} f(gk\cdot y)\,dk\,, \qquad y \in X,$$

where $g \in G$ is an element such that $g\cdot o = x$. Clearly, $F_x(y)$ is the average of f on the sphere with center x, passing through $g\cdot y$. The function F_x satisfies the decay condition (i) and it is radial. Moreover,

$$\hat{F}_x(\xi) = \int_{H_x} \hat{f}(gk\cdot\xi)\,dk. \tag{10}$$

We now need the following estimate

(11) $$d(o,gk\cdot\xi) \geq d(o,\xi) - d(o,g\cdot o).$$

For this let x_o be a point on ξ closest to $k^{-1}g^{-1}\cdot o$. Then by the triangle inequality

$$d(o,gk\cdot\xi) = d(k^{-1}g^{-1}\cdot o,\xi) \geq d(o,x_o) - d(o,k^{-1}g^{-1}\cdot o)$$
$$\geq d(o,\xi) - d(o,g\cdot o).$$

Thus it follows by (ii) that

$$\hat{F}_x(\xi) = 0 \quad \text{if} \quad d(o,\xi) > d(o,x) + R.$$

Since F_x is radial this implies by the first part of the proof that

(12) $$\int_{H_X} f(gk\cdot y)\,dk = 0$$

if

(13) $$d(o,y) > d(o,g\cdot o) + R.$$

But the set $\{gk\cdot y : k\in H_X\}$ is the sphere $S^{d(o,y)}(g\cdot o)$ with center $g\cdot o$ and radius $d(o,y)$; furthermore, the inequality in (13) implies the inclusion relation

(14) $$B^R(o) \subset B^{d(o,y)}(g\cdot o)$$

for the balls. But considering the part in $B^R(o)$ of the geodesic through o and $g\cdot o$ we see that conversely relation (14) implies (13). Theorem 1.2 will therefore be proved if we establish the following lemma.

LEMMA 1.4. _Let_ $f\in C(X)$ _satisfy the conditions_:

(i) _For each integer_ $m>0$, $f(x)e^{md(o,x)}$ _is bounded._

(ii) <u>There exists a number</u> $R > 0$ <u>such that the surface integral</u>

$$\int_S f(s)\,d\omega(s) = 0 ,$$

<u>whenever the sphere</u> S <u>encloses the ball</u> $B^R(o)$.

<u>Then</u>

$$f(x) = 0 \quad \underline{\text{for}} \quad d(o,x) > R.$$

<u>Proof</u>. This lemma is the exact analog of Lemma 2.7, Ch. I, whose proof, however, used the vector space structure of $\mathbb{R}^n$. By using a special model of the hyperbolic space we shall nevertheless adapt the proof to the present situation. As before we may assume f is smooth, i.e. $f \in C^\infty(X)$.

Consider the unit ball $\{x \in \mathbb{R}^n : \sum_1^n x_i^2 < 1\}$ with the Riemannian structure

$$(15) \qquad ds^2 = \rho(x_1, \ldots, x_n)^2(dx_1^2 + \ldots + dx_n^2)$$

where

$$\rho(x_1, \ldots, x_n) = 2(1 - x_1^2 - \ldots - x_n^2)^{-1} .$$

This Riemannian manifold is well known to have constant curvature -1 so we can use it for a model of X. This model is useful here because the spheres in X are the ordinary Euclidean spheres inside the ball. This fact is obvious for the spheres Σ with center 0. For the general statement it suffices to prove that if T is the geodesic symmetry with respect to a point (which we can take on the x_1-axis) then $T(\Sigma)$ is a Euclidean sphere. The unit disk D in the x_1x_2-plane is totally geodesic in X, hence invariant under T. Now the isometries of the non-Euclidean disk D are generated by the complex conjugation $x_1+ix_2 \to x_1-ix_2$ and fractional linear transformations so they map Euclidean circles into Euclidean circles. In particular $T(\Sigma \cap D) = T(\Sigma) \cap D$ is a Euclidean circle. But T commutes with the rotations around the x_1-axis. Thus $T(\Sigma)$ is invariant under such rotations and intersects D in a circle; hence it is a Euclidean sphere.

After these preliminaries we pass to the proof of Lemma 1.4. Let $S = S^r(y)$ be a sphere in X enclosing $B^R(o)$ and let $B^r(y)$ denote the corresponding ball. Expressing the exterior $X - B^r(y)$ as a union of spheres in X with center y we deduce from assumptions (ii)

$$\int_{B^r(y)} f(x)\,dx = \int_X f(x)\,dx \,, \tag{16}$$

which is a constant for small variations in r and y. The Riemannian measure dx is given by

$$dx = \rho^n dx_o, \tag{17}$$

where $dx_o = dx_1 \ldots dx_n$ is the Euclidean volume element. Let r_o and y_o, respectively, denote the Euclidean radius and Euclidean center of $S^r(y)$. Then $S^{r_o}(y_o) = S^r(y)$, $B^{r_o}(y_o) = B^r(y)$ set-theoretically and by (16) and (17)

$$\int_{B^{r_o}(y_o)} f(x_o)\, \rho(x_o)^n dx_o = \text{const.}\,, \tag{18}$$

for small variations in r_o and y_o; thus by differentiation with respect to r_o,

(19) $$\int_{S^{r_o}(y_o)} f(s_o)\,\rho(s_o)^n\,d\omega_o(s_o) = 0\ ,$$

where $d\omega_o$ is the Euclidean surface element. Putting $f^*(x) = f(x)\,\rho(x)^n$ we have by (18)

$$\int_{B^{r_o}(y_o)} f^*(x_o)\,dx_o = \text{Const.},$$

so, differentiating with respect to y_o, we get

$$\int_{B^{r_o}(0)} (\partial_i f^*)(y_o+x_o)\,dx_o = 0\ .$$

Using the divergence theorem (26), §2 , on the vector field $F(x_o) = f^*(y_o+x_o)\partial_i$ defined in a neighborhood of $B^{r_o}(0)$ the last equation implies

$$\int_{S^{r_o}(0)} f^*(y_o+s)s_i\,d\omega_o(s) = 0$$

which in combination with (19) gives

(20) $$\int_{S^{r_o}(y_o)} f^*(s)s_i\,d\omega_o(s) = 0\ ,$$

The Euclidean and the non-Euclidean Riemannian structures on $S^{r_o}(y_o)$ differ by the factor ρ^2. It follows that $d\omega = \rho(s)^{n-1}\,d\omega_o$ so (20) takes the form

(21) $$\int_{S^r(y)} f(s)\,\rho(s)s_i\,d\omega(s) = 0.$$

We have thus proved that the function $x \to f(x)\rho(x)x_i$ satisfies the assumptions of the theorem. By iteration we obtain

(22) $$\int_{S^r(y)} f(s)\,\rho(s)^k\,s_{i_1}\cdots s_{i_k}\,d\omega(s) = 0.$$

In particular, this holds with $y = 0$ and $r > R$. Then $\rho(s) = \text{constant}$ and (22) gives $f \equiv 0$ outside $B^R(o)$ by the Weierstrass approximation theorem. Now Theorem 1.2 is proved.

Now let L denote the Laplace-Beltrami operator on X. (See Ch. IV, §1 for the definition). Because of formula (3) for the Riemannian structure of X, L is given by

$$L = \frac{\partial^2}{\partial r^2} + (n-1)\coth r \frac{\partial}{\partial r} + (\sinh r)^{-2} L_S \tag{23}$$

where L_S is the Laplace-Beltrami operator on the unit sphere in X_o. We consider also for each $r \geq 0$ the mean value operator M^r defined by

$$(M^r f)(x) = \frac{1}{A(r)} \int_{S^r(x)} f(s)\, d\omega(s) .$$

As we saw before this can also be written

$$(M^r f)(g \cdot o) = \int_{H_X} f(gk \cdot y)\, dk \tag{24}$$

if $g \in G$ is arbitrary and $y \in X$ is such that $r = d(o,y)$. If f is an analytic function one can, by expanding it in a Taylor series, prove from (24) that M^r is a certain power series in L (cf. Helgason [1959], p. 270-272). In particular we have the commutativity

$$M^r L = L M^r . \tag{25}$$

This in turn implies the "Darboux equation"

$$L_x(F(x,y)) = L_y(F(x,y)) \tag{26}$$

for the function $F(x,y) = (M^{d(o,y)}f)(x)$. In fact, using (24) and (25) we have if $g\cdot o = x$, $r = d(o,y)$

$$L_x(F(x,y)) = (LM^r f)(x) = (M^r L f)(x)$$

$$= \int_{H_X} (Lf)(gk\cdot y)\,dk = \int_{H_X} (L_y(f(gk\cdot y))\,dk$$

the last equation following from the invariance of the Laplacian under the isometry gk. But this last expression is $L_y(F(x,y))$.

Remark. The analog of Lemma 2.13 in Ch. IV which also holds here would give another proof of (25) and (26).

For a fixed integer k $(1\leqslant k\leqslant n-1)$ let Ξ denote the manifold of all k-dimensional totally geodesic submanifolds of X. If ϕ is a continuous function on Ξ we denote by $\check{\phi}$ the point function

$$\check{\phi}(x) = \int_{x\in\xi} \phi(\xi)\,d\mu(\xi) \,,$$

where μ is the unique measure on the (compact) space of ξ passing through x, invariant under all rotations around x and having total measure one.

THEOREM 1.5. (*The inversion formula*). *For* k *even let* Q_k *denote the polynomial*

$$Q_k(z) = [z+(k-1)(n-k)]\,[z+(k-3)(n-k+2)]\,\ldots\,[z+1\cdot(n-2)]$$

of degree $k/2$.. *The* k-*dimensional Radon transform on* X *is then inverted by the formula*

$$c f = Q_k(L)\left((\hat{f})^{\vee}\right), \qquad f \in C_c^{\infty}(X).$$

Here c *is the constant*

(27) $$c = \frac{\Gamma(\frac{1}{2}n)}{\Gamma(\frac{1}{2}(n-k))}(-4\pi)^{\frac{1}{2}k}.$$

The formula holds also for f *rapidly decreasing in the sense of condition* (i) *in Theorem* 4.2.

Proof. Fix $\xi \in \Xi$ passing through the origin $o \in X$. If $x \in X$ fix $g \in G$ such that $g \cdot o = x$. As k runs through H_X, $gk \cdot \xi$ runs through the set of totally geodesic submanifolds of X passing through x and

$$\check{\phi}(g \cdot o) = \int_K \phi(gk \cdot \xi)\, dk .$$

Hence

$$(\hat{f})^{\vee}(g \cdot o) = \int_K \left(\int_\xi f(gk \cdot y)\, dm(y) \right) dk = \int_\xi (M^r f)(g \cdot o)\, dm(y) ,$$

where $r = d(o,y)$. But since ξ is totally geodesic in X, it has also constant curvature -1 and two points in ξ have the same distance in ξ as in X. Thus we have

(28) $$(\hat{f})^{\vee}(x) = \Omega_k \int_0^{\infty} (M^r f)(x)(\sinh r)^{k-1} dr .$$

We apply L to both sides and use (23). Then

(29) $$\left(L(\hat{f})^{\vee}\right)(x) = \Omega_k \int_0^{\infty} (\sinh r)^{k-1} L_r(M^r f)(x))\, dr ,$$

where L_r is the "radial part" $\frac{\partial^2}{\partial r^2}+(n-1)\coth r\,\frac{\partial}{\partial r}$ of L. Putting now $F(r) = (M^r f)(x)$ we have the following result.

LEMMA 1.6. *Let* m *be an integer* $0<m<n=\dim X$. *Then*

$$\int_0^\infty \sinh^m r\; L_r F\,dr$$
$$= (m+1-n)\left[m\int_0^\infty \sinh^m r\; F(r)\,dr + (m-1)\int_0^\infty \sinh^{m-2} r\,F(r)\,dr\right].$$

If $m=1$ *the term* $(m-1)\int_0^\infty \sinh^{m-2} r\,F(r)\,dr$ *should be replaced by* $F(0)$.

This follows by repeated integration by parts.

From this lemma combined with the Darboux equation (26) in the form

$$L_x(M^r f(x)) = L_r(M^r f(x)) \tag{30}$$

we deduce

$$\left[L_x+m(n-m-1)\right]\int_0^\infty \sinh^m r (M^r f)(x)\,dr$$
$$= -(n-m-1)(m-1)\int_0^\infty \sinh^{m-2} r (M^r f)(p)\,dr.$$

Applying this repeatedly to (29) we obtain Theorem 4.5.

B. *The Spheres and the Elliptic Spaces*. Now let X be the unit sphere $S^n(o)\subset\mathbb{R}^{n+1}$ and Ξ the set of k-dimensional totally geodesic submanifolds of X. Each $\xi\in\Xi$ is a k-sphere. We shall now invert the Radon transform

$$\hat{f}(\xi) = \int_\xi f(x)\,dm(x), \qquad f\in C^\infty(X)$$

where dm is the measure on ξ given by the Riemannian structure induced by that of X. In contrast to the hyperbolic space, each geodesic in X through a point x also passes through the antipodal point A_x. As a result, $\hat{f} = (f \circ A)^{\wedge}$ and our inversion formula will reflect this fact. Although we state our result for the sphere, it is really a result for the _elliptic space_, that is the sphere with antipodal points identified. The functions on this space are naturally identified with symmetric functions on the sphere.

Again let

$$\check{\phi}(x) = \int_{x \in \xi} \phi(\xi)\, d\mu(\xi)$$

denote the average of a continuous function on Ξ over the set of ξ passing through x.

THEOREM 1.7. _Let_ k _be an integer_, $1 \leq k < n = \dim X$.

(i) _The mapping_ $f \to \hat{f}$ $(f \in C^\infty(X))$ _has kernel consisting of the skew function (the functions_ f _satisfying_ $f + f \circ A = 0$).

(ii) _Assume_ k _even and let_ P_k _denote the polynomial_

$$P_k(z) = [z-(k-1)(n-k)]\,[z-(k-3)(n-k+2)]\ \ldots\ [z-1(n-2)]$$

of degree $k/2$. _The_ k-_dimensional Radon transform on_ X _is then inverted by the formula_

$$c(f + f \circ A) = P_k(L)((\hat{f})^{\vee}), \qquad f \in C^\infty(X)$$

where c _is the constant in_ (27).

Proof. We first prove (ii) in a similar way as in the noncompact case. The Riemannian structure in (3) is now replaced by

$$ds^2 = dr^2 + \sin^2 r\, d\sigma^2 \;;$$

the Laplace-Beltrami operator is now given by

$$L = \frac{\partial^2}{\partial r^2} + (n-1)\cot r \frac{\partial}{\partial r} + (\sin r)^{-2} L_S \tag{31}$$

instead of (23) and

$$(\hat{f})^{\vee}(x) = \Omega_k \int_0^{\pi} (M^r f)(x) \sin^{k-1} r \, dr.$$

For a fixed x we put $F(r) = (M^r f)(x)$. The analog of Lemma 1.6 now reads as follows.

LEMMA 1.8. *Let* m *be an integer*, $0 < m < n = \dim X$. *Then*

$$\int_0^{\pi} \sin^m r \, L_r F \, dr$$

$$= (n-m-1)\left[m \int_0^{\pi} \sin^m r \, F(r)\, dr - (m-1) \int_0^{\pi} \sin^{m-2} r \, F(r)\, dr \right]$$

If $m=1$, *the term* $(m-1)\int_0^{\pi} \sin^{m-2} r \, F(r)\, dr$ *should be replaced by* $F(0) + F(\pi)$.

Since (30) is still valid the lemma implies

$$[L_x - m(n-m-1)] \int_0^{\pi} \sin^m r (M^r f)(x)\, dr$$

$$= -(n-m-1)(m-1) \int_0^{\pi} \sin^{m-2} r (M^r f)(x)\, dr$$

and the desired inversion formula follows by iteration since $F(0) + F(\pi) = f(x) + f(Ax)$.

In the case when k is even, Part (i) follows from (ii). Next suppose $k=n-1$, n even. For each ξ there are exactly two points x and Ax at maximum distance, namely $\frac{\pi}{2}$, from ξ and we write

$$\hat{f}(x) = \hat{f}(Ax) = \hat{f}(\xi).$$

We have then

(32) $$\hat{f}(x) = \Omega_n(M^{\frac{\pi}{2}}f)(x).$$

Next we recall some well known facts about spherical harmonics. We have

(33) $$L^2(X) = \sum_0^\infty \mathcal{H}_s ,$$

where the space $\mathcal{H}_s$ consists of the restrictions to X of the homogeneous harmonic polynomials on $\mathbb{R}^{n+1}$ of degree s.

(a) $Lh_s = -s(s+n-1)h_s$ $(h_s \in \mathcal{H}_s)$ for each $s \geq 0$. This is immediate from the decomposition

$$L_{n+1} = \frac{\partial^2}{\partial r^2} + \frac{n}{r}\frac{\partial}{\partial r} + \frac{1}{r^2}L$$

of the Laplacian L_{n+1} of $\mathbb{R}^{n+1}$ (cf. (23)). Thus the spaces $\mathcal{H}_s$ are precisely the eigenspaces of L.

(b) Each $\mathcal{H}_s$ contains a function $(\not\equiv 0)$ which is invariant under the group K of rotations around the vertical axis (the x_{n+1}-axis in $\mathbb{R}^{n+1}$). This function ϕ_s is nonzero at the north pole o and is uniquely determined by the condition $\phi_s(o) = 1$. This is easily seen since by (31) ϕ_s satisfies the ordinary differential equation

$$\frac{d^2\phi_s}{dr^2} + (n-1)\cot r \frac{d\phi_s}{dr} = -s(s+n-1)\,\phi_s , \qquad \phi_s'(o) = 0$$

It follows that $\mathcal{H}_s$ is irreducible under the orthogonal group $\mathbf{O}(n+1)$.

(c) Since the mean value operator $M^{\pi/2}$ commutes with the action of $\mathbf{O}(n+1)$ it acts as a scalar c_s on the irreducible space $\mathcal{H}_s$. Since we have

$$M^{\pi/2}\phi_s = c_s\,\phi_s\,, \qquad\qquad \phi_s(o) = 1,$$

we obtain

$$c_s = \phi_s(\tfrac{\pi}{2})\,. \tag{34}$$

LEMMA 1.9. <u>The scalar</u> $\phi_s(\pi/2)$ <u>is zero if and only if</u> s <u>is odd</u>.

<u>Proof</u>. Let H_s be the K-invariant homogeneous harmonic polynomial whose restriction to X equals ϕ_s. Then H_s is a polynomial in $x_1^2+\ldots+x_n^2$ and x_{n+1} so if the degree s is odd, x_{n+1} occurs in each term whence $\phi_s(\pi/2) = H_s(1,0,\ldots,0,0) = 0$. If s is even, say $s = 2d$, we write

$$H_s = a_o(x_1^2+\ldots+x_n^2)^d + a_1x_{n+1}^2(x_1^2+\ldots+x_n^2)^{d-1}+\ldots+a_dx_{n+1}^{2d}.$$

Using $L_{n+1}=L_n+\partial^2/\partial x_{n+1}^2$ and formula (31) in Ch.I the equation $L_{n+1}H_s \equiv 0$ gives the recursion formula

$$a_i(2d-2i)(2d-2i+n-2) + a_{i+1}(2i+2)(2i+1) = 0$$

$(0\leq i<d)$. Hence $H_s(1,0\ldots,0)$, which equals a_o, is $\neq 0$; Q. e. d.

Now each $f\in C^\infty(X)$ has a uniformly convergent expansion

$$f = \sum_o^\infty h_s \qquad\qquad (h_s\in\mathcal{H}_s)$$

and by (32)

$$\hat{f} = \Omega_n \sum_0^\infty c_s h_s .$$

If $\hat{f} = 0$ then by Lemma 1.9, $h_s = 0$ for s even so f is skew. Conversely $\hat{f} = 0$ if f is skew so Theorem 1.7 is proved for the case $k=n-1$, n even.

If k is odd, $0<k<n-1$, the proof just carried out shows that $\hat{f}(\xi) = 0$ for all $\xi \in \Xi$ implies that f has integral 0 over every $(k+1)$ - dimensional sphere with radius 1 and center o. Since $k+1$ is even and $< n$ we conclude by (ii) that $f + f \circ A = 0$ so the theorem is proved.

§ 2. Compact Two-point Homogeneous Spaces. Applications.

We shall now extend the inversion formula in Theorem 1.7 to compact two-point homogeneous spaces X of dimension $n>1$. By virtue of Wang's classification [1952] these are also the compact symmetric spaces of rank one (see Matsumoto [1971] for a more direct proof) so their geometry can be described very explicitly. Here we shall use some geometric and group theoretic properties of these spaces ((i) - (vii) below) and refer to Helgason ([1959], p.278, [1965A], § 5-6 or [1978, Ch. VII §10]) for their proofs.

Let U denote the group $I(X)$ of isometries of X. Fix an origin $o \in X$ and let K denote the isotropy subgroup U_o. Let $\mathfrak{k}$ and $\mathfrak{u}$ be the Lie algebras of K and U, respectively. Then $\mathfrak{u}$ is semisimple. Let $\mathfrak{p}$ be the orthogonal complement of $\mathfrak{k}$ in $\mathfrak{u}$ with respect to the Killing form B of $\mathfrak{u}$. Changing the distance function on X by a constant factor we may assume that the differential of the mapping

$u \to u \cdot o$ of U onto X gives an isometry of $\mathfrak{p}$ (with the metric of $-B$) onto the tangent space X_o. This is the canonical metric on X which we shall use.

Let L denote the diameter of X, that is the maximal distance between any two points. If $x \in X$ let A_x denote the set of points in X of distance L from x. By the two-point homogeneity the isotropy subgroup U_x acts transitively on A_x; thus $A_x \subset X$ is a submanifold, _the antipodal manifold_ associated to x.

(i) _Each A_x is a totally geodesic submanifold of X; with the Riemannian structure induced by that of X it is another two-point homogeneous space._

(ii) _Let Ξ denote the set of all antipodal manifolds in X; since U acts transitively on Ξ, the set Ξ has a natural manifold structure. Then the mapping $j: x \to A_x$ is a one-to-one diffeomorphism; also $x \in A_y$ if and only if $y \in A_x$._

(iii) _Each geodesic in X has period $2L$. If $x \in X$ the mapping $\mathrm{Exp}_x: X_x \to X$ gives a diffeomorphism of the ball $B^L(0)$ onto the open set $X - A_x$._

Fix a vector $H \in \mathfrak{p}$ of length L (i.e. $L^2 = -B(H,H)$). For $Z \in \mathfrak{p}$ let T_Z denote the linear transformation $Y \to [Z,[Z,Y]]$ of $\mathfrak{p}$, $[\, ,\,]$ denoting the Lie bracket in $\mathfrak{u}$. For simplicity, we now write Exp instead of Exp_o. A point $Y \in \mathfrak{p}$ is said to be _conjugate_ to o if the differential $d\mathrm{Exp}$ is singular at Y.

The line $\mathfrak{a} = \mathbb{R}H$ is a maximal abelian subspace of $\mathfrak{p}$. The eigenvalues of T_H are 0, $\alpha(H)^2$ and possibly $(\frac{1}{2}\alpha(H))^2$ where $\pm\alpha$ (and possibly $\pm\frac{1}{2}\alpha$) are the roots of $\mathfrak{u}$ with respect to $\mathfrak{a}$. Let

$$\mathfrak{p} = \mathfrak{a} + \mathfrak{p}_\alpha + \mathfrak{p}_{\frac{1}{2}\alpha} \tag{35}$$

be the corresponding decomposition of $\mathfrak{p}$ into eigenspaces; the dimensions $q = \dim(\mathfrak{p}_\alpha)$, $p = \dim(\mathfrak{p}_{\frac{1}{2}\alpha})$ are called the <u>multiplicities</u> of α and $\frac{1}{2}\alpha$, respectively.

(iv) *Suppose H is conjugate to 0. Then $\operatorname{Exp}(\mathfrak{a}+\mathfrak{p}_\alpha)$, with the Riemannian structure induced by that of X, is a sphere, totally geodesic in X, having o and $\operatorname{Exp} H$ as antipodal points, and having curvature π^2/L^2. Moreover*

$$A_{\operatorname{Exp} H} = \operatorname{Exp}(\mathfrak{p}_{\frac{1}{2}\alpha}).$$

(v) *If H is not conjugate to o then $\mathfrak{p}_{\frac{1}{2}\alpha} = 0$ and*

$$A_{\operatorname{Exp} H} = \operatorname{Exp} \mathfrak{p}_\alpha.$$

(vi) *The differential at Y of Exp is given by*

$$d\operatorname{Exp}_Y = d\tau(\exp Y) \circ \sum_0^\infty \frac{T_Y{}^k}{(2k+1)!},$$

where for $u \in U$, $\tau(u)$ is the isometry $x \to u \cdot x$.

(vii) *In analogy with (23) the Laplace-Beltrami operator L on X has the expression*

$$L = \frac{\partial^2}{\partial r^2} + \frac{1}{A(r)} A'(r) \frac{\partial}{\partial r} + L_{S^r},$$

where L_{S^r} is the Laplace-Beltrami operator on $S^r(o)$ and $A(r)$ its area.

LEMMA 2.1. *The surface area $A(r)$ $(0<r<L)$ is given by*

$$A(r) = \Omega_n \lambda^{-p}(2\lambda)^{-q} \sin^p(\lambda r) \sin^q(2\lambda r)$$

where p and q are the multiplicities above and $\lambda = |\alpha(H)|/2L$.

Proof. Because of (iii) and (vi) the surface area of $S^r(o)$ is given by

$$A(r) = \int_{|Y|=r} \det\left(\sum_0^{\infty} \frac{T_Y{}^k}{(2k+1)!}\right) d\omega_r(Y) ,$$

where $d\omega_r$ is the surface element on the sphere $|Y|=r$ in $\mathfrak{p}$. Because of the two-point homogeneity the integrand depends on r only so we can evaluate it for $Y = H_r = \frac{r}{L}H$. Since the nonzero eigenvalues of T_{H_r} are $\alpha(H_r)^2$ with multiplicity q and $(\frac{1}{2}\alpha(H_r))^2$ with multiplicity p, a trivial computation gives the lemma.

We consider now Problems A, B and C from Chapter II, §2 for the homogeneous spaces X and Ξ, which are acted on transitively by the same group U. Fix an element $\xi_0 \in \Xi$ passing through the origin $o \in X$. If $\xi_0 = A_{0'}$ then an element $u \in U$ leaves ξ_0 invariant if and only if it lies in the isotropy subgroup $K' = U_{0'}$; we have the identifications

$$X = U/K \quad , \quad \Xi = U/K'$$

and $x \in X$ and $\xi \in \Xi$ are incident if and only if $x \in \xi$.

On Ξ we now choose a Riemannian structure such that the diffeomorphism $j : x \to A_x$ from (ii) is an isometry. Let L and Λ denote the Laplacians on X and Ξ, respectively. With $\check{x}$ and $\hat{\xi}$ defined as in Ch. II, §1, we have

$$\hat{\xi} = \xi \ , \quad \check{x} = \{j(y) : y \in j(x)\} \quad ;$$

the first relation amounts to the incidence description above and the second is a consequence of the property $x \in A_y \Leftrightarrow y \in A_x$ listed under (ii).

The sets $\check{x}$ and $\hat{\xi}$ will be given the measures $d\mu$ and dm, respectively, induced by the Riemannian structures of Ξ and X. The Radon transform and its dual are then given by

$$\hat{f}(\xi) = \int_{\xi} f(x)\, dm(x) \quad , \quad \check{\phi}(x) = \int_{\check{x}} \phi(\xi)\, d\mu(\xi) .$$

But

$$\check{\phi}(x) = \int_{\check{x}} \phi(\xi)\, d\mu(\xi) = \int_{y \in j(x)} \phi(j(y))\, d\mu(j(y)) = \int_{j(x)} (\phi \circ j)(y)\, dm(y)$$

so

(36) $$\check{\phi} = (\phi \circ j)^{\wedge} \circ j .$$

Because of this correspondence between the transforms $f \to \hat{f}$, $\phi \to \check{\phi}$ it suffices to consider the first one. Let $\mathbb{D}(X)$ denote the algebra of differential operators on X, invariant under U. It can be shown that $\mathbb{D}(X)$ is generated by L. Similarly $\mathbb{D}(\Xi)$ is generated by Λ.

THEOREM 2.2. (i) <u>The mapping</u> $f \to \hat{f}$ <u>is a linear one-to-one mapping of</u> $C^{\infty}(X)$ <u>onto</u> $C^{\infty}(\Xi)$ <u>and</u>

$$(Lf)^{\wedge} = \Lambda \hat{f} .$$

(ii) <u>Except for the case when</u> X <u>is an even-dimensional elliptic space</u>

$$f = P(L)((\hat{f})^{\vee}), \qquad f \in C^{\infty}(X),$$

<u>where</u> P <u>is a polynomial, independent of</u> f, <u>explicitly given below,</u> (44) - (50). <u>In all cases</u>

degree (P) = ½ dimension of the antipodal manifold.

Proof. We first prove (ii). Let dk be the Haar measure on K such that $\int dk = 1$ and let Ω_X denote the total measure of an antipodal manifold in X. Then $\mu(\check{o}) = m(A_o) = \Omega_X$ and if $u \in U$,

$$\check{\phi}(u \cdot o) = \Omega_X \int_K \phi(uk \cdot \xi_o)\, dk.$$

Hence

$$(\hat{f})^\vee(u \cdot o) = \Omega_X \int_K \left(\int_{\xi_o} f(uk \cdot y)\, dm(y) \right) dk$$

$$= \Omega_X \int_{\xi_o} (M^r f)(u \cdot o)\, dm(y),$$

where r is the distance $d(o,y)$ in the space X between o and y. If $d(o,y) < L$ there is a unique geodesic in X of length $d(o,y)$ joining o to y and since ξ_o is totally geodesic, $d(o,y)$ is also the distance in ξ_o between o and y. Thus using geodesic polar coordinates in ξ_o in the last integral we obtain

$$(\hat{f})^\vee(x) = \Omega_X \int_0^L (M^r f)(x) A_1(r)\, dr, \tag{37}$$

where $A_1(r)$ is the area of a sphere of radius r in ξ_o. By Lemma 2.1 we have

$$A_1(r) = C_1 \sin^{p_1}(\lambda_1 r) \sin^{q_1}(2\lambda_1 r), \tag{38}$$

where C_1 and λ_1 are constants and p_1, q_1 are the multiplicities for the antipodal manifold. In order to prove (ii) on the basis of

(37) we need the following complete list of the compact symmetric spaces of rank one and their corresponding antipodal manifolds:

	X		A_o
Spheres	S^n	$(n=1,2,..)$	point
Real projective spaces	$\mathbb{P}^n(\mathbb{R})$	$(n=2,3,\ldots)$	$\mathbb{P}^{n-1}(\mathbb{R})$
Complex " "	$\mathbb{P}^n(\mathbb{C})$	$(n=4,6,\ldots)$	$\mathbb{P}^{n-2}(\mathbb{C})$
Quaternian " "	$\mathbb{P}^n(\mathbb{H})$	$(n=8,12,\ldots)$	$P^{n-4}(\mathbb{H})$
Cayley plane	$\mathbb{P}^{16}(\mathbb{C}ay)$		S^8

Here the superscripts denote the real dimension. For the lowest dimensions, note that

$$\mathbb{P}^1(\mathbb{R}) = S^1, \quad \mathbb{P}^2(\mathbb{C}) = S^2, \quad \mathbb{P}^4(\mathbb{H}) = S^4.$$

For the case S^n, (ii) is trivial and the case $X = \mathbb{P}^n(\mathbb{R})$ was already done in Theorem 1.7. For the remaining cases α and $\frac{1}{2}\alpha$ are both roots so by (v) H is conjugate to o, so we have the properties in (iv). Since a closed geodesic in A_o is a closed geodesic in X we have

$$\begin{aligned} L &= \text{diameter } X = \text{diameter } A_x \\ &= \text{distance from } o \text{ to the nearest conjugate point in } X_o \\ &= \text{smallest number } M \text{ such that } \lim_{r\to M} A(r) = 0. \end{aligned}$$

The multiplicities p and q for the rank-one symmetric spaces were determined by Cartan [1927] (see also Araki [1962], Helgason [1978], p. 532) and we can now derive the following list:

$X = \mathbb{P}^n(\mathbb{C})$:

$$p = n-2, \quad q = 1, \quad \lambda = \pi/2L,$$

$$A(r) = \tfrac{1}{2}\,\Omega_n \lambda^{-n+1} \sin^{(n-2)}(\lambda r)\sin(2\lambda r),$$

$$A_1(r) = \tfrac{1}{2}\,\Omega_{n-2} \lambda^{-n+3} \sin^{(n-4)}(\lambda r)\sin(2\lambda r);$$

$X = \mathbb{P}^n(\mathbb{H})$:

$$p = n-4,\ q = 3,\ \lambda = \pi/2L,$$

$$A(r) = \frac{1}{8}\,\Omega_n \lambda^{-n+1} \sin^{(n-4)}(\lambda r)\sin^3(2\lambda r),$$

$$A_1(r) = \frac{1}{8}\,\Omega_{n-4} \lambda^{-n+5} \sin^{n-8}(\lambda r)\sin^3(2\lambda r);$$

$X = \mathbb{P}^{16}(\mathbb{C}\mathrm{ay})$:

$$p = 8,\quad q = 7,\quad \lambda = \pi/2L,$$

$$A(r) = \frac{1}{2^7}\,\Omega_{16} \lambda^{-15} \sin^8(\lambda r)\sin^7(2\lambda r),$$

$$A_1(r) = \Omega_8 \sin^7(2\lambda r).$$

In all cases do we have

$$\Omega_X = m(A_o) = \int_o^L A_1(r)\,dr. \tag{39}$$

Thus $A(r)$ and $A_1(r)$ are in all cases expressed in terms of n and L. But L can also be expressed in terms of n. In fact, the Killing form metric has the property

$$B(H,H) = \sum_\beta \beta(H)^2$$

as β runs over the roots. Thus

$$-L^2 = -|H|^2 = 2p(\tfrac{1}{2}\alpha(H))^2 + 2q(\alpha(H))^2.$$

But by (iii) H is the first conjugate point on the ray $\mathbb{R}^+H$ so $|\alpha(H)| = \pi$ (Helgason [1978], p. 294). Thus we get the formula

$$L^2 = p\,\frac{\pi^2}{2} + 2q\,\pi^2. \tag{40}$$

LEMMA 2.3. <u>In the Killing form metric the diameter</u> L <u>of the projective spaces</u>

$$\mathbb{P}^n(\mathbb{C}) , \quad \mathbb{P}^n(\mathbb{H}) , \quad \mathbb{P}^{16}(\mathbf{Cay})$$

is respectively given by

$$(\tfrac{n}{2}+1)^{\frac{1}{2}}\pi , \quad (\tfrac{n}{2}+4)^{\frac{1}{2}}\pi , \quad 3\sqrt{2}\,\pi .$$

As already used for spaces of constant curvature we have here the commutativity, cf. (25),

$$(41) \qquad M^r L = L M^r$$

which implies (cf. (26))

$$(42) \qquad L_x((M^r f)(x)) = L_r((M^r f)(x)) ,$$

where by (vii) and Lemma 2.1,

$$L_r = \frac{\partial^2}{\partial r^2} + \lambda\{p \cot(\lambda r) + 2q \cot(2\lambda r)\} \frac{\partial}{\partial r} , \qquad (0 < r < L) .$$

We now apply the Laplacian to (37) and use (42) in order to simplify the right-hand side. The result is reduced by repeated use of the following three lemmas.

LEMMA 2.4. _Let_ $X = \mathbb{P}^n(\mathbb{C})$, $f \in C^\infty(X)$. _If_ m _is an even integer,_ $0 \leq m \leq n-4$ _then_

$$\Big(L_x - \lambda^2 (n-m-2)(m+2)\Big) \int_0^L \sin^m(\lambda r) \sin(2\lambda r)[M^r f](x)\, dr$$
$$= -\lambda^2 (n-m-2)m \int_0^L \sin^{m-2}(\lambda r) \sin(2\lambda r)[M^r f](x)\, dr .$$

For $m = 0$ _the right-hand side should be replaced by_

$$-2\lambda(n-2) f(x) .$$

LEMMA 2.5. _Let_ $X = \mathbb{P}^n(\mathbb{H})$, $f \in C^\infty(X)$. _Let_ m _be an even integer,_

$0 < m \leq n-8$. Then

$$(L_x - \lambda^2 (n-m-4)(m+6)) \int_0^L \sin^m(\lambda r)\sin^3(2\lambda r)\left[M^r f\right](x)\,dr$$

$$= -\lambda^2 (n-m-4)(m+2) \int_0^L \sin^{m-2}(\lambda r)\sin^3(2\lambda r)\left[M^r f\right](x)\,dr .$$

Also

$$\left[(L_x - 6\lambda^2(n-4))(L_x - 4\lambda^2(n-2))\right]\left(\int_0^L \sin^3(2\lambda r)\left[M^r f\right](x)\,dr\right) = 16\lambda^3(n-2)(n-4)f(x).$$

LEMMA 2.6. Let $X = \mathbb{P}^{16}(\mathbb{C}ay)$, $f \in C^\infty(X)$. Let $m > 1$ be an integer. Then

$$(L_x - 4\lambda^2 m(11-m)) \int_0^L \sin^m(2\lambda r)\left[M^r f\right](x)\,dr$$

$$= -32\lambda^2 (m-1) \int_0^L \sin^{m-2}(2\lambda r)\cos^2(\lambda r)\left[M^r f\right](x)$$

$$+ 4\lambda^2 (m-1)(m-7) \int_0^L \sin^{m-2}(2\lambda r)\left[M^r f\right](x)\,dr ;$$

$$(L_x - 4\lambda^2 (m+1)(10-m)) \int_0^L \sin^m(2\lambda r)\cos^2(\lambda r)\left[M^r f\right](x)\,dr$$

$$= 4\lambda^2 (3m-5) \int_0^L \sin^m(2\lambda r)\left[M^r f\right](x)\,dr$$

$$+4\lambda^2 (m-1)(m-15) \int_0^L \sin^{m-2}(2\lambda r)\cos^2(\lambda r)\left[M^r f\right](x)\,dr .$$

Also

(43) $$(L_x - 72\lambda^2) \int_0^L \sin(2\lambda r)\cos^2(\lambda r)(M^r f)(x)\,dr$$

$$= -8\lambda^2 \int_0^L \sin(2\lambda r)(M^r f)(r)\,dr - 28\lambda\, f(x).$$

These lemmas are proved by means of long computations. Since the methods are similar for all cases let us just verify the last formula (43). Here we have

$$L_r = \frac{\partial^2}{\partial r^2} + \lambda\{8\cot(\lambda r) + 14\cot(2\lambda r)\}\frac{\partial}{\partial r}.$$

So, putting $F(r) = (M^r f)(x)$, we have by (42)

$$L_X \int_0^L \sin(2\lambda r)\cos^2(\lambda r)(M^r f)(x)\,dr =$$

$$\int_0^L \left[\sin(2\lambda r)\cos^2(\lambda r)F''(r) + (44\lambda\cos^4(\lambda r) - 14\lambda\cos^2(\lambda r))F'(r)\right]dr$$

$$= \int_0^L \left[36\lambda\cos^4(\lambda r) - 8\lambda\cos^2(\lambda r)\right]F'(r)\,dr$$

$$= -28\lambda F(o) - \lambda^2 \int_0^L F(r)\left[\sin(2\lambda r)(8 - 72\cos^2(\lambda r))\right]dr$$

which gives formula (43).

We can now prove (ii) in Theorem 2.2. Consider first the case $X = \mathbb{P}^{16}(\mathrm{Cay})$. We have

$$(\hat{f})^{\vee}(x) = \Omega_8\,\Omega_X \int_0^L (M^r f)(x)\sin^7(2\lambda r)\,dr.$$

Here

$$L^2 = 18\pi^2, \quad \Omega_X = \Omega_8 \int_0^L \sin^7(2\lambda r)\,dr$$

$$\lambda = \pi/2L.$$

Taking $m = 7$ in Lemma 2.6 we get

$$(L_X - 112\lambda^2)\int_0^L (M^r f)(x)\sin^7(2\lambda r)\,dr$$

$$= -192\lambda^2 \int_0^L (M^r f)(x)\sin^5(2\lambda r)\cos^2(\lambda r)\,dr$$

and then taking $m = 5$ we get

$$(L_x - 120\,\lambda^2)(L_x - 112\,\lambda^2)\int_0^L (M^r f)(x)\sin^7(2\lambda r)\,dr$$

$$= (-192\,\lambda^2)(40\lambda^2)\left[\int_0^L (M^r f)(x)\sin^5(2\lambda r)\,dr\right.$$

$$\left. - 4\int_0^L (M^r f)(x)\sin^3(2\lambda r)\cos^2(\lambda r)\,dr\right].$$

Taking $m = 5$ again we get

$$(L_x - 120\,\lambda^2)^2 (L_x - 112\,\lambda^2)\left[\int_0^L (M^r f)(x)\sin^7(2\lambda r)\,dr\right]$$

$$= (-192\,\lambda^2)(40\lambda^2)\left[(-128\,\lambda^2)\int_0^L \sin^3(2\lambda r)\cos^2(\lambda r)(M^r f)(x)\,dr\right.$$

$$- 32\lambda^2\int_0^L \sin^3(2\lambda r)(M^r f)(x)\,dr$$

$$\left. - 4(L_x - 120\,\lambda^2)\int_0^L (M^r f)(x)\sin^3(2\lambda r)\cos^2(\lambda r)\,dr\right].$$

Taking $m = 3$ the last term is found to be

$$- 4(L_x - 112\,\lambda^2 - 8\,\lambda^2)\int_0^L (M^r f)(x)\sin^3(2\lambda r)\cos^2(\lambda r)\,dr$$

$$= - 64\,\lambda^2\int_0^L (M^r f)(x)\sin^3(2\lambda r)$$

$$+ 384\,\lambda^2\int_0^L (M^r f)(x)\sin(2\lambda r)\cos^2(\lambda r)\,dr$$

$$+ 32\lambda^2\int_0^L (M^r f)(x)\sin^3(2\lambda r)\cos^2(\lambda r)\,dr\,.$$

Hence

$$(L_x - 120\,\lambda^2)^2\,(L_x - 112\,\lambda^2)\int_0^L (M^r f)(x)\sin^7(2\lambda r)\,dr$$

$$= 192 \cdot 40 \cdot 96\, \lambda^6 \int_0^L (M^r f)(x) \sin^3(2\lambda r) \cos^2(\lambda r)\, dr$$

$$+ \int_0^L (M^r f)(x) \sin^3(2\lambda r)\, dr$$

$$- 4 \int_0^L (M^r f)(x) \sin(2\lambda r) \cos^2(\lambda r)\, dr \ .$$

Finally we apply the operator $(L_x - 112\,\lambda^2)$ to both sides. Taking $m = 3$ in Lemma 2.6 we get

$$(L_x - 112\,\lambda^2) \int_0^L (M^r f)(x) \sin^3(2\lambda r) \cos^2(\lambda r)\, dr$$

$$= 16\,\lambda^2 \int_0^L (M^r f)(x) \sin^3(2\lambda r)\, dr$$

$$- 96\,\lambda^2 \int_0^L (M^r f)(x) \sin(2\lambda r) \cos^2(\lambda r)\, dr$$

$$(L_x - 112\,\lambda^2) \int_0^L (M^r f)(x) \sin^3(2\lambda r)\, dr$$

$$= -16\,\lambda^2 \int_0^L (M^r f)(x) \sin^3(2\lambda r)\, dr$$

$$-64\,\lambda^2 \int_0^L (M^r f)(x) \sin(2\lambda r) \cos^2(\lambda r)\, dr$$

$$-32\,\lambda^2 \int_0^L (M^r f)(x) \sin(2\lambda r)\, dr$$

$$-4(L_x - 112\,\lambda^2) \int_0^L (M^r f)(x) \sin(2\lambda r) \cos^2(\lambda r)\, dr$$

$$= 160\,\lambda^2 \int_0^L (M^r f)(x) \sin(2\lambda r) \cos^2(\lambda r)\, dr$$

$$+ 32\lambda^2 \int_0^L (M^r f)(x) \sin(2\lambda r)\, dr$$

$$+ 112\lambda\, f(x).$$

Fortunately, all terms except the last one cancel out and we obtain

$$(L_x - 112\lambda^2)^2 (L_x - 120\lambda^2)^2 \int_0^L (M^r f)(x) \sin^7(2\lambda r)\, dr$$

$$= 192 \cdot 40 \cdot 96 \cdot 112 \cdot \lambda^7 f(x).$$

If we now substitute the values $\lambda^{-2} = 72$ we get

$$\left(L_x - \frac{14}{9}\right)^2 \left(L_x - \frac{15}{9}\right)^2 (\hat{f})^\vee (x)$$

$$= \Omega_8^2 \frac{1}{2\lambda} \left(\int_0^\pi \sin^7 s\, ds\right) 192 \cdot 40 \cdot 96 \cdot 112\, \lambda^7 f(x)$$

$$= \frac{\pi^8}{9} \left(\int_0^\pi \sin^7 s\, ds\right) 2^8 \cdot 3^{-4} \cdot 5 \cdot 7\, f(x) = \frac{\pi^8 2^{13}}{3^6} f(x).$$

Thus we have proved for $X = \mathbb{P}^{16}(\mathbb{C}ay)$,

$$f = P(L)((\hat{f})^\vee), \qquad f \in C^\infty(X),$$

where

$$P(L) = \frac{3^6}{\pi^8 2^{13}} \left(L - \frac{14}{9}\right)^2 \left(L - \frac{15}{9}\right)^2. \tag{44}$$

For $X = \mathbb{P}^n(\mathbb{C})$ we find similarly from Lemma 2.4 the formula

$$f = P(L)((\hat{f})^\vee), \qquad f \in C^\infty(X)$$

where, since $\lambda^{-2} = 2(n+2)$,

$$(45)\qquad P(L) = c\left(L - \frac{(n-2)2}{2(n+2)}\right)\left(L - \frac{(n-4)4}{2(n+2)}\right)\cdots\left(L - \frac{2(n-2)}{2(n+2)}\right)$$

with

$$(46)\qquad c = \left(-8\pi^2(n+2)\right)^{1-\frac{n}{2}} .$$

For $X = \mathbb{P}^n(\mathbb{H})$ we derive from Lemma 2.5 the formula

$$f = P(L)((\hat{f})^\vee) \qquad\qquad f \in C^\infty(X)$$

where, since $\lambda^{-2} = 2(n+8)$,

$$(47)\qquad P(L) = c\left(L - \frac{(n-2)4}{2(n+8)}\right)\left(L - \frac{(n-4)6}{2(n+8)}\right)\cdots\left(L - \frac{4(n-2)}{2(n+8)}\right)$$

with

$$(48)\qquad c = \tfrac{1}{2}\left(-4\pi^2(n+8)\right)^{2-\frac{n}{2}} .$$

Finally we determine $P(L)$ for the case $X = \mathbb{P}^n(\mathbb{R})$ because now the metric on X is normalized by means of the Killing form of $U = I(X)$ rather than by the curvature $+1$ condition in Theorem 1.7. Instead of the functions on $\mathbb{P}^n(\mathbb{R})$ we shall deal with even functions f on the sphere $\mathbf{S}^n$ and define $\hat{f}(\xi)$ as the integral of f over the totally geodesic $(n-1)$-sphere Ξ. In order to preserve (36) we define $\check{\phi}(x)$ as an integral over an $(n-1)$-sphere (of unit normal vectors to hyperplanes through x).

The Killing form metric on $\mathbf{S}^n$ is obtained by multiplying the usual Riemannian structure (curvature $+1$) by $2n$ (see Helgason [1961]). The new Laplacian is therefore $\frac{1}{2n}$ times the Laplacian in Theorem 1.7. Thus we have by Theorem 1.7, with $k = n-1$, n odd,

$$P(L)((\hat{f})^{\vee}) = f ,$$

where

$$(49) \qquad P(L) = c\left(L - \frac{(n-2)1}{2n}\right)\left(L - \frac{(n-4)3}{2n}\right) \ . \ . \ . \ \left(L - \frac{1(n-2)}{2n}\right) ,$$

c a constant. To find c we apply the formula to the function $f \equiv 1$. Since

$$\hat{1} \equiv \Omega_n (2n)^{\frac{1}{2}(n-1)} , \quad (\hat{1})^{\vee} \equiv \Omega_n{}^2 (2n)^{n-1}$$

we obtain

$$(50) \qquad c = \frac{1}{4}(-4\pi^2 n)^{\frac{1}{2}(1-n)} .$$

This concludes the proof of Part (ii) of Theorem 2.2.

That $f \to \hat{f}$ is injective follows from (ii) except for the case $X = \mathbb{P}^n(\mathbb{R})$, n even. But in this exceptional case the injectivity follows from Theorem 1.7.

For the surjectivity we use once more the fact that the mean-value operator M^r commutes with the Laplacian (cf. (41)). We have

$$(51) \qquad \hat{f}(j(x)) = c(M^L f)(x) ,$$

where c is a constant. Thus by (36)

$$(\hat{f})^{\vee}(x) = (\hat{f} \circ j)^{\wedge}(j(x)) = c M^L (\hat{f} \circ j)(x)$$

so

$$(52) \qquad (\hat{f})^{\vee} = c^2 M^L M^L f .$$

Thus if X is not an even-dimensional projective space f is a constant multiple of $M^L P(L) M^L f$ which by (51) shows $f \to \hat{f}$ surjective. For the case $X = \mathbb{P}^n(\mathbb{R})$, n even, we use Theorem 1.7. If the map

$f \to \hat{f}$ were not surjective there would by (51) be a nonzero distribution T on X such that

$$T(M^L f) = 0 \qquad f \in C^\infty(X). \tag{53}$$

Now take f to be an eigenfunction of L; then, as used earlier, f is an eigenfunction of M^L and, by the injectivity, with a nonzero eigenvalue. Thus (53) implies the contradiction $T = 0$.

It remains to prove $(Lf)^\wedge = \Lambda\hat{f}$. For this we use (36), (41) and (51). By the definition of Λ we have

$$(\Lambda\phi)(j(x)) = L(\phi \circ j)(x) \qquad x \in X,\ \phi \in C^\infty(\Xi).$$

Thus

$$(\Lambda\hat{f})(j(x)) = (L(\hat{f} \circ j))(x) = cL(M^L f)(x) = cM^L(L f)(x) = (Lf)^\wedge(j(x)).$$

This finishes the proof of Theorem 2.2.

COROLLARY 2.7. <u>Let</u> X <u>be a compact two-point homogeneous space and suppose</u> f <u>satisfies</u>

$$\int_\gamma f(x)\, ds(x) = 0$$

<u>for each (closed) geodesic</u> γ <u>in</u> X, ds <u>being the element of arc-length</u>. <u>Then</u>

(i) <u>If</u> X <u>is a sphere</u>, f <u>is skew</u>.

(ii) <u>If</u> X <u>is not a sphere</u>, $f \equiv 0$.

Taking a convolution with f we may assume f smooth. Part (i) is already contained in Theorem 1.7. For Part (ii) we use the classification; for $X = \mathbb{P}^{16}(\mathbf{Cay})$ the antipodal manifolds are totally geodesic spheres so using Part (i) we conclude that $\hat{f} \equiv 0$ so by Theorem 2.2, $f \equiv 0$. For the remaining cases $\mathbb{P}^n(\mathbb{C})$ $(n = 4, 6, \ldots)$ and $\mathbb{P}^n(\mathbb{H})$, $(n = 8, 12, \ldots)$ (ii) follows similarly by induction as the initial antipodal manifolds, $\mathbb{P}^2(\mathbb{C})$ and $\mathbb{P}^4(\mathbb{H})$, are totally geodesic spheres.

COROLLARY 2.8. *Let* B *be a bounded open set in* $\mathbb{R}^{n+1}$, *symmetric and star-shaped with repect to* 0, *bounded by a hypersurface. Assume for a fixed* k $(1 \leq k < n)$

$$\text{Area}\,(B \cap P) = \text{constant} \tag{54}$$

for all $(k+1)$*-planes* P *through* 0. *Then* B *is an open ball*.

In fact, we know from Theorem 1.7 that if f is a symmetric function on $X = \mathbf{S}^n$ with $\hat{f}(\mathbf{S}^n \cap P)$ constant (for all P) then f is a constant. We apply this to the function

$$f(\theta) = \rho(\theta)^{k+1} \qquad \theta \in \mathbf{S}^n$$

if $\rho(\theta)$ is the distance from the origin to each of the two points of intersection of the boundary of B with the line through 0 and θ; f is well defined since B is symmetric. If $\theta = (\theta_1, \ldots, \theta_k)$ runs through the k-sphere $\mathbf{S}^n \cap P$ then the point

$$x = \theta r \qquad (0 \leq r < \rho(\theta))$$

runs through the set $B \cap P$ and

$$\text{Area}(B \cap P) = \int_{\mathbf{S}^n \cap P} d\omega(\theta) \int_0^{\rho(\theta)} r^k\, dr .$$

It follows that $\text{Area}(B \cap P)$ is a constant multiple of $\hat{f}(\mathbf{S}^n \cap P)$ so (54) implies that f is constant. This proves the corollary.

§ 3. Noncompact Two-point Homogeneous Spaces.

Theorem 2.2 has an analog for noncompct two-point homogeneous spaces which we shall now describe. By Tits' classification [1955] p.183 of homogeneous manifolds L/H for which L act transitively on the tangents to L/H it is known, in principle, what the noncompact two-point homogeneous spaces are. As in the compact case they turn out to be symmetric. A direct proof of this fact was given by Nagano [1959] and Helgason [1959]. The theory of symmetric spaces then implies that the noncompact two-point homogeneous spaces are the Euclidean spaces and the noncompact spaces $X = G/K$ where G is a connected semisimple Lie group with finite center and real rank one and K a maximal compact subgroup.

Let $\mathfrak{g} = \mathfrak{k} + \mathfrak{p}$ be the direct decomposition of the Lie algebra of G into the Lie algebra $\mathfrak{k}$ of K and its orthogonal complement $\mathfrak{p}$ (with respect to the Killing form of $\mathfrak{g}$). Fix a one-dimensional subspace $\mathfrak{a} \subset \mathfrak{p}$ and let

$$\mathfrak{p} = \mathfrak{a} + \mathfrak{p}_\alpha + \mathfrak{p}_{\frac{1}{2}\alpha} \tag{55}$$

be the decomposition of $\mathfrak{p}$ into eigenspaces of T_H (in analogy with (35)). Let ξ_0 denote the totally geodesic submanifold $\operatorname{Exp}(\mathfrak{p}_{\frac{1}{2}\alpha})$; in the case $\mathfrak{p}_{\frac{1}{2}\alpha} = 0$ we put $\xi_0 = \operatorname{Exp}(\mathfrak{p}_\alpha)$. By the classification and duality for symmetric spaces we have the following complete list of the spaces G/K. In the list the superscript denotes the real dimension; for the lowest dimensions note that

$$\mathbb{H}^1(\mathbb{R}) = \mathbb{R}, \quad \mathbb{H}^2(\mathbb{C}) = \mathbb{H}^2(\mathbb{R}), \quad \mathbb{H}^4(\mathbb{H}) = \mathbb{H}^4(\mathbb{R}).$$

X		ξ_0
Real Hyperbolic spaces	$\mathbb{H}^n(\mathbb{R})$ $(n=2,3,\ldots)$,	$\mathbb{H}^{n-1}(\mathbb{R})$
Complex " "	$\mathbb{H}^n(\mathbb{C})$ $(n=4,6,\ldots)$,	$\mathbb{H}^{n-2}(\mathbb{C})$
Quaternian " "	$\mathbb{H}^n(\mathbb{H})$ $(n=8,12,\ldots)$,	$\mathbb{H}^{n-4}(\mathbb{H})$
Cayley " "	$\mathbb{H}^{16}(\mathbb{C}\mathrm{ay})$,	$\mathbb{H}^8(\mathbb{R})$.

Let Ξ denote the set of submanifolds $g\cdot\xi_0$ of X as g runs through G; Ξ is given the canonical differentiable structure of a homogeneous space. Each $\xi \in \Xi$ has a measure m induced by the Riemannian structure of X and the Radon transform on X is defined by

$$\hat{f}(\xi) = \int_\xi f(x)\, dm(x), \qquad f \in C_c(X).$$

The dual transform $\phi \to \check{\phi}$ is defined by

$$\check{\phi}(x) = \int_{\xi \ni x} \phi(\xi)\, d\mu(\xi), \qquad \phi \in C(\Xi),$$

where μ is the invariant average on the set of ξ passing through x. Let L denote the Laplace-Beltrami operator on X, the Riemannian structure being that given by the Killing form of $\mathfrak{g}$.

THEOREM 3.1. *The Radon transform $f \to \hat{f}$ is a one-to-one mapping of $C_c^\infty(X)$ into $C_c^\infty(\Xi)$ and, except for the case $X = \mathbb{H}^n(\mathbb{R})$, n even, is inverted by the formula*

$$f = Q(L)((\hat{f})^\vee),$$

the polynomial Q being given as follows:

$X = \mathbb{H}^n(\mathbb{R})$, n odd:

$$Q(L) = \gamma\left(L+\frac{(n-2)1}{2n}\right)\left(L+\frac{(n-4)3}{2n}\right)\ \cdots\ \left(L+\frac{1(n-2)}{2n}\right).$$

$X = \mathbb{H}^n(\mathbb{C})$:

$$Q(L) = \gamma\left(L+\frac{(n-2)2}{2(n+2)}\right)\left(L+\frac{(n-4)4}{2(n+2)}\right)\ \cdots\ \left(L+\frac{2(n-2)}{2(n+2)}\right).$$

$X = \mathbb{H}^n(\mathbb{H})$:

$$Q(L) = \gamma\left(L+\frac{(n-2)4}{2(n+8)}\right)\left(L+\frac{(n-4)6}{2(n+8)}\right)\cdots\left(L+\frac{4(n-2)}{2(n+8)}\right).$$

$X = \mathbb{H}^{16}(\mathbf{Cay})$:

$$Q(L) = \gamma\left(L+\frac{14}{9}\right)^2\left(L+\frac{15}{9}\right)^2.$$

The constants γ are obtained from the constants c in (44), (46), (48) _and_ (50) _by multiplication by the factor Ω_X in_ (39).

We omit the proof since it is quite analogous to that of Theorem 2.2. The adjustment from the constants c to the constants γ is based on the difference in the normalizations of the dual Radon transform in the compact and the noncompact case.

§ 4. The X-Ray Transform on a Symmetric Space.

Let X be a complete Riemannian manifold of dimension > 1 in which any two points can be joined by a unique geodesic. The _X-ray transform_ on X assigns to each continuous function f on X the integrals

$$\hat{f}(\gamma) = \int_\gamma f(x)\, ds(x), \tag{56}$$

γ being any complete geodesic in X and ds the element of arc-length. In analogy with the X-ray reconstruction problem on $\mathbb{R}^n$ (Ch. I, §7) one can consider the problem of inverting the X-ray transform $f \to \hat{f}$. With d denoting the distance in X and $o \in X$ some fixed point we now define two subspaces of $C(X)$. Let

$$F(X) = \{f \in C(X): \sup_x d(o,x)^k |f(x)| < \infty \text{ for each } k \geq 0\}$$

$$\mathcal{F}(X) = \{f \in C(X): \sup_x e^{kd(o,x)} |f(x)| < \infty \text{ for each } k \geq 0\}$$

Because of the triangle inequality these spaces do not depend on the choice of o. We can informally refer to $F(X)$ as the space of continuous _rapidly decreasing functions_ and to $\mathcal{F}(X)$ as the space of continuous _exponentially decreasing functions_. We shall now prove the analog of the support theorems (Theorem 2.6, Ch. I, Theorem 1.2, Ch.III) for the X-ray transform on a symmetric space of the noncompact type. This general analog turns out to be a direct corollary of the Euclidean case and the hyperbolic case, already done.

COROLLARY 4.1. _Let_ X _be a symmetric space of the noncompact type_, B _any ball in_ M.

(i) _If a function_ $f \in \mathcal{F}(X)$ _satisfies_

(57) $$\hat{f}(\gamma) = 0 \qquad \text{whenever } \gamma \cap B = \emptyset$$

then

(58) $$f(x) = 0 \qquad \text{for } x \notin B.$$

In particular, the X-ray transform is one-to-one on $\mathcal{F}(X)$.

(ii) _If_ X _has rank greater than one statement_ (i) _holds with_ $\mathcal{F}(X)$ _replaced by_ $F(X)$.

Proof. Let o be the center of B, r its radius, and let γ be an arbitrary geodesic in X through o.

Assume first X has rank greater than one. By a standard conjugacy theorem for symmetric spaces γ lies in a two-dimensional, flat, totally geodesic submanifold of X. Using Theorem 2.6, Ch.I on this Euclidean plane we deduce $f(x) = 0$ if $x \in \gamma$, $d(o,x) > r$. Since γ is arbitrary (58) follows.

Next suppose X has rank one. Identifying $\mathfrak{p}$ with the tangent space X_o let $\mathfrak{a}$ be the tangent line to γ. We can then consider the eigenspace decomposition (55). If $\mathfrak{b} \subset \mathfrak{p}_\alpha$ is a line through the origin then $S = \operatorname{Exp}(\mathfrak{a}+\mathfrak{b})$ is a totally geodesic submanifold of X (cf. (iv) in the beginning of §2). Being 2-dimensional and not flat, S is necessarily a hyperbolic space. From Theorem 1.2 we therefore conclude $f(x) = 0$ for $x \in \gamma$, $d(o,x) > r$. Again (58) follows since γ is arbitrary.

BIBLIOGRAPHICAL NOTES

It was shown by Funk [1916] that a function f on the 2-sphere, symmetric with respect to the center, can be determined by the integrals of f over the great circles. In [1917] Radon discussed this problem and the analogous one of determining a function on the non-Euclidean plane from its integrals over all geodesics; he also indicated corresponding inversion formulas.

The Radon transform on hyperbolic and on elliptic spaces corresponding to totally geodesic submanifolds was defined in Helgason [1959] where the inversion formulas in Theorems 1.5 and 1.7 were proved. A generalization was given by Semyanistyi [1961]. An alternative definition, with corresponding inversion formulas, was given in Gelfand-Graev-Vilenkin [1962].

The support theorem (Theorem 1.2) was proved by the author ([1964 A] , [1980 A]) and its consequence, Cor. 4.1, pointed out in [1980 B]. The theory of the Radon transform for antipodal manifolds in compact two-point homogeneous spaces (Theorem 2.2) is from Helgason [1965 A]. R. Michel has in [1972] and [1973] used Theorem 2.2 in establishing certain infinitesimal rigidity properties of the canonical metrics on the real and complex projective spaces. See also Guillemin [1976] and A. Besse [1978].

CHAPTER IV

ORBITAL INTEGRALS AND THE WAVE OPERATOR FOR ISOTROPIC LORENTZ SPACES.

In Chapter II, § 3 we discussed the problem of determining a function on a homogeneous space by means of its integrals over generalized spheres. We shall now solve this problem for the isotropic Lorentz spaces (Theorem 4.1 below). As we shall presently explain these spaces are the Lorentzian analogs of the two-point homogeneous spaces considered in Chapter III.

§ 1. Isotropic Spaces.

Let X be a manifold. A pseudo-Riemannian structure of signature (p,q) is a smooth assignment $y \to g_y$ where g_y is a symmetric non-degenerate bilinear form on $X_y \times X_y$ of signature (p,q). This means that for a suitable basis $Y_1, \ldots, Y_{p+q}$ of X_y we have

$$g_y(Y) = y_1^{\,2} + \ldots + y_p^{\,2} - y_{p+1}^{\,2} - \ldots - y_{p+q}^{\,2}$$

if $Y = \sum_1^{p+q} y_i Y_i$. If $q=0$ we speak of a Riemannian structure and if $p=1$ we speak of a Lorentzian structure. Connected manifolds X with such structures g are called pseudo-Riemannian (respectively Riemannian, Lorentzian) manifolds.

A manifold X with a pseudo-Riemannian structure g has a differential operator of particular interest, the so-called Laplace-Beltrami operator. Let $(x_1, \ldots, x_{p+q})$ be a coordinate system on an open subset U of X. We define the functions g_{ij}, g^{ij}, and $\bar{g}$ on U by

$$g_{ij} = g\left(\frac{\partial}{\partial x_i}, \frac{\partial}{\partial x_j}\right), \quad \sum_j g_{ij} g^{jk} = \delta_{ik},$$

$$\bar{g} = |\det(g_{ij})| .$$

The Laplace-Beltrami operator L is defined on U by

$$Lf = \frac{1}{\sqrt{\bar{g}}}\left(\sum_k \frac{\partial}{\partial x_k}\left(\sum_i g^{ik}\sqrt{\bar{g}}\,\frac{\partial f}{\partial x_i}\right)\right)$$

for $f \in C^\infty(U)$. It is well known that this expression is invariant under coordinate changes so L is a differential operator on X.

An isometry of a pseudo-Riemannian manifold X is a diffeomorphism preserving g. It is easy to prove (see e.g. Helgason [1959] p. 245) that L is invariant under each isometry ϕ, that is $L(f \circ \phi) = (Lf) \circ \phi$ for each $f \in C^\infty(X)$. Let $I(X)$ denote the group of all isometries of X. For $y \in X$ let $I(X)_y$ denote the subgroup of $I(X)$ fixing y (the isotropy subgroup at y) and let H_y denote the group of linear transformations of the tangent space X_y induced by the action of $I(X)_y$. For each $a \in \mathbb{R}$ let $\Sigma_a(y)$ denote the "sphere"

(1) $$\Sigma_a(y) = \{ Z \in X_y : g_y(Z,Z) = a, \quad Z \neq 0 \} .$$

Definition. The pseudo-Riemannian manifold X is called *isotropic* if for each $a \in \mathbb{R}$ and each $y \in X$ the group H_y acts transitively on $\Sigma_a(y)$.

PROPOSITION 1.1. *An isotropic pseudo-Riemannian manifold X is homogeneous; that is, $I(X)$ acts transitively on X.*

Proof. The pseudo-Riemannian structure on X gives an affine connection preserved by each isometry $g \in I(X)$. Any two points $y, z \in X$ can be joined by a curve consisting of finitely many geodesic segments γ_i $(1 \leq i \leq p)$. Let g_i be an isometry fixing the midpoint of γ_i and reversing the tangents to γ_i at this point. The product $g_p \cdots g_1$ maps y to z, whence the homogeneity of X.

A. The Riemannian Case.

The following simple result shows that the isotropic spaces are natural generalizations of the spaces considered in the last chapter.

PROPOSITION 1.2. *A Riemannian manifold X is isotropic if and only if it is two-point homogeneous.*

Proof. If X is two-point homogeneous and $y \in X$ the isotropy subgroup $I(X)_y$ at y is transitive on each sphere $S^r(y)$ in X with center y so X is clearly isotropic. On the other hand if X is isotropic it is homogeneous (Prop. 1.1) hence complete; thus by standard Riemannian geometry any two points in X can be joined by means of a geodesic. Now the isotropy of X implies that for each $y \in X$, $r > 0$ the group $I(X)_y$ is transitive on the sphere $S^r(y)$, whence the two-point homogeneity.

B. The General pseudo-Riemannian Case.

Let X be a manifold with a pseudo-Riemannian structure g and curvature tensor R. Let $y \in X$, $S \subset X_y$ a two-dimensional subspace on which g_y is nondegenerate. The curvature of X along the section S is defined by

$$K(S) = -\frac{g_p(R_p(X,Y)X,Y)}{g_p(X,X)g_p(Y,Y) - g_p(X,Y)^2}$$

The denominator is in fact $\neq 0$ and the expression is independent of the choice of X and Y.

We shall now construct isotropic pseudo-Riemannian manifolds of signature (p,q) and constant curvature. Consider the space $\mathbb{R}^{p+q+1}$ with the flat pseudo-Riemannian structure

$$B_e(Y) = y_1^2 + \dots + y_p^2 - y_{p+1}^2 - \dots - y_{p+q}^2 + e y_{p+q+1}^2, \quad (e = \pm 1).$$

Let Q_e denote the quadric in $\mathbb{R}^{p+q+1}$ given by

$$B_e(Y) = e.$$

The orthogonal group $\mathbf{O}(B_e)$ $(= \mathbf{O}(p,q+1)$ or $\mathbf{O}(p+1,q))$ acts transitively on Q_e; the isotropy subgroup at $o = (0, \dots, 0, 1)$ is identified with $\mathbf{O}(p,q)$.

THEOREM 1.3. (i) _The restriction of_ B_e _to the tangent spaces to_ Q_e _gives a pseudo-Riemannian structure_ g_e _on_ Q_e _of signature_ (p,q).

(ii) _We have_

$$(2) \qquad Q_{-1} \cong \mathbf{O}(p,q+1)/\mathbf{O}(p,q) \qquad \text{(diffeomorphism)}$$

and the pseudo-Riemannian structure g_{-1} _on_ Q_{-1} _has constant curvature_ -1.

(iii) _We have_

$$(3) \qquad Q_{+1} = \mathbf{O}(p+1,q)/\mathbf{O}(p,q) \qquad \text{(diffeomorphism)}$$

and the pseudo-Riemannian structure g_{+1} _on_ Q_{+1} _has constant curvature_ +1.

(iv) _The flat space_ $\mathbb{R}^{p+q}$ _with the quadratic form_

$g_0(Y) = \sum_1^p y_i^2 - \sum_{p+1}^{p+q} y_j^2$ *and the spaces*

$$\mathbf{O}(p,q+1)/\mathbf{O}(p,q) \ , \ \mathbf{O}(p+1,q)/\mathbf{O}(p,q)$$

are all isotropic and (up to a constant factor on the pseudo-Riemannian structure) exhaust the class of pseudo-Riemannian manifolds of constant curvature and signature (p,q) *except for local isometry*.

Proof. If s_o denotes the linear transformation

$$(y_1,\ldots,y_{p+q}, y_{p+q+1}) \rightarrow (-y_1,\ldots,-y_{p+q}, y_{p+q+1})$$

then the mapping $\sigma : g \to s_o g s_o$ is an involutive automorphism of $\mathbf{O}(p,q+1)$ whose differential $d\sigma$ has fixed point set $\mathfrak{o}(p,q)$ (the Lie algebra of $\mathbf{O}(p,q)$). The (-1)-eigenspace of $d\sigma$, say $\mathfrak{m}$, is spanned by the vectors

$$(4) \qquad Y_i = E_{i,p+q+1} + E_{p+q+1,i} \qquad (1 \leq i \leq p),$$

$$(5) \qquad Y_j = E_{j,p+q+1} - E_{p+q+1,j} \qquad (p+1 \leq j \leq p+q).$$

Here E_{ij} denotes a square matrix with entry 1 where the ith row and the jth column meet, all other entries being 0.

The mapping $\psi : g\,\mathbf{O}(p,q) \to g \cdot o$ has a differential $d\psi$ which maps $\mathfrak{m}$ bijectively onto the tangent plane $y_{p+q+1} = 1$ to Q_{-1} at o and $d\psi(X) = X \cdot o$ $(X \in \mathfrak{m})$. Thus

$$d\psi(Y_k) = (\delta_{1k},\ldots, \delta_{p+q+1,k}), \qquad (1 \leq k \leq p+q).$$

Thus $B_{-1}(d\psi(Y_k)) = 1$ if $1 \leq k \leq p$ and -1 if $p+1 \leq k \leq p+q$, proving (i). Next, since the space (2) is symmetric its curvature tensor satisfies

$$R_o(X,Y)(Z) = -[[X,Y],Z],$$

where $[,]$ is the Lie bracket. A simple computation then shows

$$K(\mathbb{R}Y_k + \mathbb{R}Y_\ell) = -1 \qquad (1 \leq k, \ell \leq p+q)$$

and this implies (ii). Part (iii) is proved in the same way. For (iv) we first verify that the spaces listed are isotropic. Since the isotropy action of $\mathbf{O}(p,q+1)_o = \mathbf{O}(p,q)$ on $\mathfrak{m}$ is the ordinary action of $\mathbf{O}(p,q)$ on $\mathbb{R}^{p+q}$ it suffices to verify that $\mathbb{R}^{p+q}$ with the quadratic form g_o is isotropic. But we know $\mathbf{O}(p,q)$ is transitive on $g_e = +1$ and on $g_e = -1$ so it remains to show $\mathbf{O}(p,q)$ transitive on the cone $\{Y \neq 0 : g_e(Y) = 0\}$. By rotation in $\mathbb{R}^p$ and in $\mathbb{R}^q$ it suffices to verify the statement for $p = q = 1$. But for this case it is obvious. The uniqueness in (iv) follows from the general fact that a symmetric space is determined locally by its pseudo-Riemannian structure and curvature tensor at a point (see e.g. Helgason [1978], p.200-201). This finishes the proof.

The spaces (2) and (3) are the pseudo-Riemannian analogs of the spaces $\mathbf{O}(p,1)/\mathbf{O}(p)$, $\mathbf{O}(p+1)/\mathbf{O}(p)$ from Ch. III, §1. But the other two-point homogeneous spaces listed in Ch. III, §2 - §3 have similar pseudo-Riemannian analogs (indefinite elliptic and hyperbolic spaces over $\mathbf{C}$, $\mathbb{H}$ and $\mathbf{Cay}$). As proved by Wolf [1967], p.384, each non-flat isotropic pseudo-Riemannian manifold is locally isometric to one of these models.

We shall later need a lemma about the connectivity of the groups $\mathbf{O}(p,q)$. Let $I_{p,q}$ denote the diagonal matrix (d_{ij}) with $d_{ii} = 1$ $(1 \leq i \leq p)$ $d_{jj} = -1$ $(p+1 \leq j \leq p+q)$ so a matrix g with transpose ${}^t g$

belongs to $\mathbb{O}(p,q)$ if and only if

$$
{}^t g\, I_{p,q}\, g = I_{p,q}. \tag{6}
$$

If $y \in \mathbb{R}^{p+q}$ let $y^T = (y_1,\dots,y_p, 0 \dots 0)$, $y^S = (0,\dots,0,y_{p+1},\dots,y_{p+q})$ and for $g \in \mathbb{O}(p,q)$ let g_T and g_S denote the matrices

$$
(g_T)_{ij} = g_{ij} \qquad (1 \le i,j \le p),
$$

$$
(g_S)_{k\ell} = g_{k\ell} \qquad (p+1 \le k,\ell \le p+q).
$$

If $g_1,\dots,g_{p+q}$ denote the column vectors of the matrix g then (6) means for the scalar products

$$
g_i^T \cdot g_i^T - g_i^S \cdot g_i^S = 1, \qquad 1 \le i \le p,
$$

$$
g_j^T \cdot g_j^T - g_j^S \cdot g_j^S = -1, \qquad p+1 \le j \le p+q,
$$

$$
g_j^T \cdot g_k^T = g_j^S \cdot g_k^S, \qquad j \ne k.
$$

LEMMA 1.4. <u>We have for each</u> $g \in \mathbb{O}(p,q)$

$$
|\det(g_T)| \ge 1 \quad , \quad |\det(g_S)| \ge 1.
$$

<u>The components of</u> $\mathbb{O}(p,q)$ <u>are obtained by</u>

(7) $\det g_T \ge 1$, $\det g_S \ge 1$; (<u>identity component</u>)

(8) $\det g_T \le -1$, $\det g_S \ge 1$;

(9) $\det g_T \ge 1$, $\det g_S \le -1$;

(10) $\det g_T \le -1$, $\det g_S \le -1$.

Thus $\mathbb{O}(p,q)$ *has* 4 *components if* $p \geq 1$, $q \geq 1$, 2 *components if* p *or* $q = 0$.

Proof. Consider the Gram determinant

$$\det \begin{pmatrix} g_1^T \cdot g_1^T & g_1^T \cdot g_2^T & \cdots & g_1^T \cdot g_p^T \\ g_2^T \cdot g_1^T & & \cdot & \\ \vdots & & & \\ g_p^T \cdot g_1^T, & \cdots & & g_p^T \cdot g_p^T \end{pmatrix},$$

which equals $(\det g_T)^2$. Using the relations above it can also be written

$$\det \begin{pmatrix} 1 + g_1^S \cdot g_1^S & g_1^S \cdot g_2^S & \cdots & g_1^S \cdot g_p^S \\ g_2^S \cdot g_1^S & & \cdot \quad \cdots & \\ \vdots & & & \\ g_p^S \cdot g_1^S & & & 1 + g_p^S \cdot g_p^S \end{pmatrix}$$

which equals 1 plus a sum of lower order Gram determinants each of which is still positive. Thus $(\det g_T)^2 \geq 1$ and similarly $(\det g_S)^2 \geq 1$. Assuming now $p \geq 1, q \geq 1$ consider the decomposition of $\mathbb{O}(p,q)$ into the four pieces (7), (8), (9), (10). Each of these is $\neq \emptyset$ because (8) is obtained from (7) by multiplication by $I_{1,p+q-1}$ etc. On the other hand, since the functions $g \to \det(g_T)$, $g \to \det(g_S)$ are continuous on $\mathbb{O}(p,q)$ the four pieces above belong to different components of $\mathbb{O}(p,q)$. But by Chevalley [1946], p.201, $\mathbb{O}(p,q)$ is homeomorphic to the product of $\mathbb{O}(p,q) \cap \mathbb{U}(p+q)$ with a Euclidean space. Since $\mathbb{O}(p,q) \cap \mathbb{U}(p+q) = \mathbb{O}(p,q) \cap \mathbb{O}(p+q)$ is homeomorphic to $\mathbb{O}(p) \times \mathbb{O}(q)$ it just remains to remark that $\mathbb{O}(n)$ has two components.

C. The Lorentzian Case .

The isotropic Lorentzian manifolds are more restricted than one might at first think on the basis of the Riemannian case. In fact there is a theorem of Lichnerowicz and Walker [1945] according to which a harmonic Lorentzian manifold has constant curvature. Since an isotropic Lorentzian manifold is harmonic we can deduce the following result from Theorem 1.3.

THEOREM 1.5. *Let X be an isotropic Lorentzian manifold (signature $(1,q)$, $q \geq 1$). Then X has constant curvature so (after a multiplication of the Lorentzian structure by a positive constant) X is locally isometric to one of the following:*

$$\mathbb{R}^{1+q} \qquad (\text{flat, signature } (1,q)),$$

$$Q_{-1} = \mathbf{O}(1,q+1)/\mathbf{O}(1,q) \; : \; y_1^2 - y_2^2 - \ldots - y_{q+2}^2 = -1$$

$$Q_{+1} = \mathbf{O}(2,q)/\mathbf{O}(1,q) \; : \; y_1^2 - y_2^2 - \ldots - y_{q+1}^2 + y_{q+2}^2 = 1,$$

the Lorentzian structure being induced by $y_1^2 - y_2^2 - \ldots \mp y_{q+2}^2$.

§2. Orbital Integrals.

The orbital integrals for isotropic Lorentzian manifolds are analogs to the spherical averaging operator M^r considered in Ch. I, §1, and Ch. III, §1. We start with some geometric preparation.

For manifolds X with a Lorentzian structure g we adopt the following customary terminology: If $y \in X$ the cone

$$C_y = \{ Y \in X_y \; : \; g_y(Y,Y) = 0 \}$$

is called the <u>null cone</u> (or the <u>light cone</u>) in X_y with vertex y. A nonzero vector $Y \in X_y$ is said to be <u>timelike</u>, <u>isotropic</u>, or <u>spacelike</u> if $g_y(Y,Y)$ is positive, 0, or negative, respectively. Similar designations apply to geodesics according to the type of their tangent vectors.

While the geodesics in $\mathbb{R}^{1+q}$ are just the straight lines, the geodesics in Q_{-1} and Q_{+1} can be found by the method of Ch. III, §1.

PROPOSITION 2.1. *The geodesics in the Lorentzian quadrics* Q_{-1} *and* Q_{+1} *have the following properties:*

(i) *The geodesics are the nonempty intersections of the quadrics with two-planes in* $\mathbb{R}^{2+q}$ *through the origin.*

(ii) *For* Q_{-1} *the spacelike geodesics are closed, for* Q_{+1} *the timelike geodesics are closed.*

(iii) *The isotropic geodesics are certain straight lines in* $\mathbb{R}^{2+q}$.

Proof. Part (i) follows by the symmetry considerations in Ch.III, §1. For Part (ii) consider the intersection of Q_{-1} with the two-plane $y_1 = y_4 = \ldots = y_{q+2} = 0$. The intersection is the circle $y_2 = \cos t$, $y_3 = \sin t$ whose tangent vector $(0, -\sin t, \cos t, 0, \ldots, 0)$ is clearly spacelike. Since $\mathbf{O}(1,q+1)$ permutes the spacelike geodesics transitively the first statement in (ii) follows. For Q_{+1} we intersect similarly with the two-plane $y_2 = \ldots = y_{q+1} = 0$. For (iii) we note that the two-plane $\mathbb{R}(1,0,\ldots,0,1) + \mathbb{R}(0,1,\ldots,0)$ intersects Q_{-1} in a pair of straight lines $y_1 = t$, $y_2 \pm 1$, $y_3 = \ldots = y_{q+1} = 0$, $y_{q+2} = t$ which clearly are isotropic. The transitivity of $\mathbf{O}(1,q+1)$ on the set of isotropic geodesics then implies that each of these is a straight line. The argument for Q_{+1} is similar.

LEMMA 2.2. The quadrics Q_{-1} and Q_{+1} $(q \geq 1)$ are connected.

Proof. The q-sphere being connected the point $(y_1, \ldots, y_{q+2})$ on $Q_{\mp 1}$ can be moved continuously on $Q_{\mp 1}$ to the point

$$(y_1, (y_2^2 + \ldots + y_{q+1}^2)^{\frac{1}{2}}, 0, \ldots, 0, y_{q+2})$$

so the statement follows from the fact that the hyperboloids $y_1^2 - y_2^2 \mp y_3^2 = \mp 1$ are connected.

LEMMA 2.3. The identity components of $\mathbf{O}(1,q+1)$ and $\mathbf{O}(2,q)$ act transitively on Q_{-1} and Q_{+1}, respectively, and the isotropy subgroups are connected.

Proof. The first statement comes from the general fact (see e.g. Helgason [1978], pp.121-124) that when a separable Lie group acts transitively on a connected manifold then so does its identity component. For the isotropy groups we use the description (7) of the identity component. This shows quickly that

$$\mathbf{O}_o(1,q+1) \cap \mathbf{O}(1,q) = \mathbf{O}_o(1,q),$$

$$\mathbf{O}_o(2.q) \cap \mathbf{O}(1,q) = \mathbf{O}_o(1,q)$$

the subscript o denoting identity component. Thus we have

$$Q_{-1} = \mathbf{O}_o(1,q+1)/\mathbf{O}_o(1,q),$$

$$Q_{+1} = \mathbf{O}_o(2,q)/\mathbf{O}_o(1,q),$$

proving the lemma.

We now write the spaces in Theorem 1.5 in the form $X = G/H$ where

$H = \mathbf{O}_o(1,q)$ and G is either $G^o = \mathbb{R}^{1+q} \cdot \mathbf{O}_o(1,q)$ (semi-direct product) $G^- = \mathbf{O}_o(1,q+1)$ or $G^+ = \mathbf{O}_o(2,q)$. Let o denote the origin $\{H\}$ in X that is

$$o = (0,\dots,0) \qquad \text{if } X = \mathbb{R}^{1+q}$$

$$o = (0,\dots,0,1) \qquad \text{if } X = Q_{-1} \text{ or } Q_{+1}.$$

In the cases $X = Q_{-1}$, $X = Q_{+1}$ the tangent space X_o is the hyperplane $\{y_1,\dots,y_{q+1},1\} \subset \mathbb{R}^{2+q}$.

The timelike vectors at o fill up the "interior" $\overset{o}{C}_o$ of the cone C_o. The set $\overset{o}{C}_o$ consists of two components. The component which contains the timelike vector

$$v_o = (-1,0,\dots,0)\ ,\ (-1,0,\dots,0,1)\ ,\ (-1,0,\dots,0,1)$$

in the cases G^o/H, G^-/H, G^+/H, respectively, will be called the <u>retrograde cone</u> in X_o. It will be denoted by D_o. The component of the hyperboloid $g_o(Y,Y) = r^2$ which lies in D_o will be denoted $S^r(o)$. If y is any other point of X we define C_y, D_y, $S^r(y) \subset X_y$ by $C_y = g \cdot C_o$, $D_y = g \cdot D_o$, $S^r(y) = g \cdot S^r(o)$ if $g \in G$ is chosen such that $g \cdot o = y$. This is a valid definition because the connectedness of H implies that $h \cdot D_o \subset D_o$. We also define

$$B^r(y) = \{ Y \in D_y \ :\ 0 < g_y(Y,Y) < r^2 \}.$$

If Exp denotes the exponential mapping of X_y into X, mapping rays through 0 onto geodesics through y we put

$$\overline{D}_y = \operatorname{Exp} D_y\,, \qquad \overline{C}_y = \operatorname{Exp} C_y$$

$$\overline{S}^r(y) = \operatorname{Exp} S^r(y)\,, \quad \overline{B}^r(y) = \operatorname{Exp} B^r(y)\,.$$

Again $\overline{C}_y$ and $\overline{D}_y$ are respectively called the <u>light cone</u> and <u>retrograde cone</u> in X with vertex y. For the spaces $X = Q_+$ we always assume $r < \pi$ in order that Exp will be one-to-one on $B^r(y)$ in view of Prop. 2.1 (ii).

LEMMA 2.4. <u>The negative of the Lorentzian structure on</u> $X = G/H$ <u>induces on each</u> $\overline{S}^r(y)$ <u>a Riemannian structure of constant negative curvature</u> $(q > 1)$.

<u>Proof</u>. The manifold X being isotropic the group $H = \mathbf{O}_o(1,q)$ acts transitively on $\overline{S}^r(o)$. The subgroup leaving fixed the geodesic from o with tangent vector v_o is $\mathbf{O}_o(q)$. This implies the lemma.

LEMMA 2.5. <u>The timelike geodesics from</u> y <u>intersect</u> $\overline{S}^r(y)$ <u>under a right angle.</u>

<u>Proof</u>. By the group invariance it suffices to prove this for $y = o$ and the geodesic with tangent vector v_o. For this case the statement is obvious.

Let $\tau(g)$ denote the translation $xH \to gxH$ on G/H and for $Y \in \mathfrak{m}$ let T_Y denote the linear transformation $Z \to [Y,[Y,Z]]$ of $\mathfrak{m}$ into itself. As usual, we identify $\mathfrak{m}$ with $(G/H)_o$.

LEMMA 2.6. <u>The exponential mapping</u> $\text{Exp} : \mathfrak{m} \to G/H$ <u>has differential</u>

$$d\text{Exp}_Y = d\tau(\exp Y) \circ \sum_0^\infty \frac{T_Y{}^n}{(2n+1)!} \qquad (Y \in \mathfrak{m}).$$

For the proof see Helgason [1978], p.215.

LEMMA 2.7. <u>The linear transformation</u>

$$A_Y = \sum_0^\infty \frac{T_Y^{\,n}}{(2n+1)!}$$

<u>has determinant given by</u>

$$\det A_Y = \left\{\frac{\sinh (g(Y,Y))^{\frac12}}{(g(Y,Y))^{\frac12}}\right\}^q \qquad \text{for } Q_{-1}$$

$$\det A_Y = \left\{\frac{\sin (g(Y,Y))^{\frac12}}{(g(Y,Y))^{\frac12}}\right\}^q \qquad \text{for } Q_{+1}$$

<u>for</u> Y <u>timelike</u>.

<u>Proof</u>. Consider the case of Q_{-1}. Since $\det(A_Y)$ is invariant under H it suffices to verify this for $Y = cY_1$ in (4), where $c \in \mathbb{R}$. We have $c^2 = g(Y,Y)$ and $T_{Y_1}(Y_j) = Y_j$ $(2 \leq j \leq q+1)$. Thus T_Y has the eigenvalue 0 and $g(Y,Y)$; the latter is a q-tuple eigenvalue. This implies the formula for the determinant. The case Q_{+1} is treated in the same way.

From this lemma and the description of the geodesics in Prop. 2.1 we can now conclude the following result.

PROPOSITION 2.8. (i) <u>The mapping</u> $\mathrm{Exp}: \mathfrak{m} \to Q_{-1}$ <u>is a diffeomorphism of</u> D_o <u>onto</u> $\overline{D}_o$.

(ii) <u>The mapping</u> $\mathrm{Exp}: \mathfrak{m} \to Q_{+1}$ <u>is a diffeomorphism of</u> $B^{\pi}(o)$ <u>onto</u> $\overline{B}^{\pi}(o)$.

Let dh denote a bi-invariant measure on the unimodular group H. Let $u \in C_c^\infty(X)$, $y \in X$ and $r > 0$. Select $g \in G$ such that $g \cdot o = y$ and select $x \in S^r(o)$. Consider the integral

$$\int_H u(gh \cdot x)\, dh .$$

Since the subgroup $K \subset H$ leaving x fixed is compact it is easy to see that the set

$$C_{g,x} = \{h \in H : gh \cdot x \in \text{support } (u)\}$$

is compact; thus the integral above converges. By the bi-invariance of dh it is independent of the choice of g (satisfying $g \cdot o = y$) and of the choice of $x \in S^r(o)$. In analogy with the Riemannian case (Ch. III, § 1) we thus define the operator M^r (the orbital integral) by

$$(M^r u)(y) = \int_H u(gh \cdot x)\, dh . \tag{11}$$

If g and x run through suitable compact neighborhoods, the sets $C_{g,x}$ are enclosed in a fixed compact subset of H so $(M^r u)(y)$ depends smoothly on both r and y. It is also clear from (11) that the operator M^r is invariant under the action of G: if $l \in G$ and $\tau(l)$ denotes the transformation $nH \to lnH$ of G/H onto itself then

$$M^r(u \circ \tau(l)) = (M^r u) \circ \tau(l).$$

If dk denotes the normalized Haar measure on K we have by standard invariant integration

$$\int_H u(h \cdot x)\, dh = \int_{H/K} d\dot{h} \int_K u(hk \cdot x)\, dk = \int_{H/K} u(h \cdot x)\, d\dot{h},$$

where $d\dot{h}$ is an H-invariant measure on H/K. But if $d\overline{\omega}^r$ is the volume element on $\overline{S}^r(o)$ (cf. Lemma 2.4) we have by the uniqueness of H-invariant measures on the space $H/K \approx \overline{S}^r(o)$ that

$$\int_H u(h \cdot x)\, dh = \frac{1}{A(r)} \int_{\overline{S}^r(o)} u(z)\, d\overline{\omega}^r(z), \tag{12}$$

where $A(r)$ is a positive scalar. But since g is an isometry we deduce from (12) that

$$(M^r u)(y) = \frac{1}{A(r)} \int_{\bar{S}^r(y)} u(z)\, d\bar{\omega}^r(z).$$

Now we have to determine $A(r)$.

LEMMA 2.9. <u>For a suitable fixed normalization of the Haar measure</u> dh <u>on</u> H <u>we have</u>

$$A(r) = r^q, \quad (\sinh r)^q, \quad (\sin r)^q$$

<u>for the cases</u>

$$\mathbb{R}^{1+q}, \quad \mathbf{O}(1,q+1)/\mathbf{O}(1,q), \quad \mathbf{O}(2,q)/\mathbf{O}(1,q),$$

<u>respectively</u>.

<u>Proof</u>. The relations above show that $dh = A(r)^{-1} d\bar{\omega}^r\, dk$. The mapping $\mathrm{Exp} : D_o \to \bar{D}_o$ preserves length on the geodesics through o and maps $S^r(o)$ onto $\bar{S}^r(o)$. Thus if $z \in S^r(o)$ and Z denotes the vector from 0 to z in X_o the ratio of the volume elements of $\bar{S}^r(o)$ and $S^r(o)$ at z is given by $\det(d\mathrm{Exp}_Z)$. Because of Lemmas 2.6 - 2.7 this equals

$$1, \quad \left(\frac{\sinh r}{r}\right)^q, \quad \left(\frac{\sin r}{r}\right)^q$$

for the three respective cases. But the volume element $d\omega^r$ on $S^r(o)$ equals $r^q d\omega^1$. Thus we can write in the three respective cases

$$dh = \frac{r^q}{A(r)}\, d\omega^1\, dk, \quad \frac{\sinh^q r}{A(r)}\, d\omega^1 dk, \quad \frac{\sin^q r}{A(r)}\, d\omega^1 dk.$$

But we can once for all normalize dh by $dh = d\omega^1 dk$ and for this choice our formulas for $A(r)$ hold.

Let $\square$ denote the <u>wave operator</u> on $X = G/H$, that is the Laplace-Beltrami operator for the Lorentzian structure g.

LEMMA 2.10. *Let* $y \in X$. *On the retrograde cone* $\overline{D}_y$ *the wave operator* $\square$ *can be written*

$$\square = \frac{\partial^2}{\partial r^2} + \frac{1}{A(r)} \frac{dA}{dr} \frac{\partial}{\partial r} - L_{\overline{S}^r(y)},$$

where $L_{\overline{S}^r(y)}$ *is the Laplace-Beltrami operator on* $\overline{S}^r(y)$.

Proof. We can take $y = o$. If $(\theta_1, \ldots, \theta_q)$ are coordinates on the "sphere" $S^1(o)$ in the flat space X_o then $(r\theta_1, \ldots, r\theta_q)$ are coordinates on $S^r(o)$. The Lorentzian structure on D_o is therefore given by

$$dr^2 - r^2 d\theta^2,$$

where $d\theta^2$ is the Riemannian structure of $S^1(o)$. Since A_Y in Lemma 2.7 is a diagonal matrix with eigenvalues 1 and $r^{-1}A(r)^{1/q}$ (q - times) it follows from Lemma 2.6 that the image $\overline{S}^r(o) = \mathrm{Exp}(S^r(o))$ has Riemannian structure $r^2 d\theta^2$, $\sinh^2 r\, d\theta^2$ and $\sin^2 r\, d\theta^2$ in the cases $\mathbb{R}^{1+q}$, Q_{-1}, and Q_{+1}, respectively. By the perpendicularity in Lemma 2.5 it follows that the Lorentzian structure on $\overline{D}_o$ is given by

$$dr^2 - r^2 d\theta^2, \quad dr^2 - \sinh^2 r\, d\theta^2, \quad dr^2 - \sin^2 r\, d\theta^2$$

in the three respective cases. Now the lemma follows immediately.

The operator M^r is of course the Lorentzian analog to the spherical mean value operator for isotropic Riemannian manifolds. We shall now prove that in analogy to the Riemannian case (cf. (41), Ch.III) the operator M^r commutes with the wave operator $\square$.

THEOREM 2.11. *For each of the isotropic Lorentz spaces* $X = G^-/H$, G^+/H *or* G^o/H *the wave operator* $\square$ *and the orbital integral* M^r *commute*:

$$\square M^r u = M^r \square u \qquad \textit{for} \quad u \in C_c^\infty(X).$$

(*For* G^+/H *we assume* $r < \pi$).

Given a function u on G/H we define the function $\tilde{u}$ on G by $\tilde{u}(g) = u(g \cdot o)$.

LEMMA 2.12. *There exists a differential operator* $\tilde{\square}$ *on* G *invariant under all left and all right translations such that*

$$\tilde{\square}\tilde{u} = (\square u)^{\sim} \qquad \textit{for} \quad u \in C_c^\infty(X).$$

Proof. We consider first the case $X = G^-/H$. The bilinear form

$$K(Y,Z) = \tfrac{1}{2}\mathrm{Tr}\,(YZ)$$

on the Lie algebra $\mathfrak{o}(1,q+1)$ of G^- is nondegenerate; in fact K is nondegenerate on the complexification $\mathfrak{o}(q+2,\mathbb{C})$ consisting of all complex skew symmetric matrices of order $q+2$. A simple computation shows that in the notation of (4) and (5)

$$K(Y_1,Y_1) = 1\ , \quad K(Y_j,Y_j) = -1 \qquad (2 \leq j \leq q+1).$$

Since K is symmetric and nondegenerate there exists a unique left invariant pseudo-Riemannian structure $\tilde{K}$ on G^- such that $\tilde{K}_e = K$. Moreover, since K is invariant under the conjugation $Y \to gYg^{-1}$ of $\mathfrak{o}(1,q+1)$, $\tilde{K}$ is also right invariant. Let $\tilde{\square}$ denote the corresponding

Laplace-Beltrami operator on G^-. Then $\tilde{\square}$ is invariant under all left and right translations on G^-. Let $u \in C_c^\infty(X)$. Since $\tilde{\square}\tilde{u}$ is invariant under all right translations from H there is a unique function $v \in C^\infty(X)$ such that $\tilde{\square}\tilde{u} = \tilde{v}$. The mapping $u \to v$ is a differential operator which at the origin must coincide with $\square$, that is $\tilde{\square}\,\tilde{u}(e) = \square u(o)$. Since, in addition, both $\square$ and the operator $u \to v$ are invariant under the action of G^- on X it follows that they coincide. This proves $\tilde{\square}\tilde{u} = (\square u)^\sim$.

The case $X = G^+/H$ is handled in the same manner. For the flat case $X = G^o/H$ let

$$Y_j = (0, \ldots, 1, \ldots, 0),$$

the jth coordinate vector on $\mathbb{R}^{1+q}$. Then $\square = Y_1^2 - Y_2^2 - \cdots - Y_{q+1}^2$. Since $\mathbb{R}^{1+q}$ is naturally imbedded in the Lie algebra of G^o we can extend Y_j to a left invariant vector field $\tilde{Y}_j$ on G^o. The operator $\tilde{\square} = \tilde{Y}_1^2 - \tilde{Y}_2^2 - \cdots - \tilde{Y}_{q+1}^2$ is then a left and right invariant differential operator on G^o and again we have $\tilde{\square}\tilde{u} = (\square u)^\sim$. This proves the lemma.

We can now prove Theorem 2.11. If $g \in G$ let $L(g)$ and $R(g)$, respectively, denote the left and right translations $l \to gl$, and $l \to lg$ on G. If $l \cdot o = x$, $x \in S^r(o)$ $(r > 0)$ and $g \cdot o = y$ then

$$(M^r u)(y) = \int_H \tilde{u}(ghl)\, dh$$

because of (11). As g and l run through sufficiently small compact neighborhoods the integration takes place within a fixed compact subset of H as remarked earlier. Denoting by subscript the argument on which a differential operator is to act we shall prove the following result.

LEMMA 2.13.

$$\tilde{\square}_1\left(\int_H \tilde{u}(ghl)\,dh\right)=\int_H (\tilde{\square}\tilde{u})(ghl)\,dh=\tilde{\square}_g\left(\int_H \tilde{u}(ghl)\,dh\right).$$

Proof. The first equality sign follows from the left invariance of $\tilde{\square}$. In fact, the integral on the left is

$$\int_H (\tilde{u}\circ L(gh))(1)\,dh$$

so

$$\tilde{\square}_1\left(\int_H \tilde{u}(ghl)\,dh\right) = \int_H\left[\tilde{\square}(\tilde{u}\circ L(gh))\right](1)\,dh$$

$$=\int_H\left[(\tilde{\square}\tilde{u})\circ L(gh)\right](1)\,dh = \int_H (\tilde{\square}\tilde{u})(ghl)\,dh\,.$$

The second equality in the lemma follows similarly from the right invariance of $\tilde{\square}$. But this second equality is just the commutativity statement in Theorem 2.11.

Lemma 2.13 also implies the following analog of the Darboux equation in Lemma 3.2, Ch. I.

COROLLARY 2.14. Let $u\in C_c^\infty(X)$ and put

$$U(y,z) = (M^r u)(y) \qquad \text{if } z\in S^r(o).$$

Then

$$\square_y(U(y,z)) = \square_z(U(y,z)).$$

REMARK 2.15. In $\mathbb{R}^n$ the solutions to the Laplace equation $Lu=0$ are characterized by the spherical mean-value theorem $M^r u=u$ (all r). This can be stated equivalently: $M^r u$ is a constant in r. In this latter form the mean value theorem holds for the solutions of the wave equation $\square u=0$ in an isotropic Lorentzian manifold: If u satisfies

$\Box u=0$ and if u is suitably small at ∞ then $(M^r u)(o)$ is constant in r. For a precise statement and proof see Helgason [1959], p.289.

§ 3. Generalized Riesz Potentials.

In this section we generalize part of the theory of Riesz potentials (Ch.I, §8) to isotropic Lorentz spaces.

Consider first the case

$$X = Q_{-1} = G^{-}/H = \mathbf{O}_o(1,n)/\mathbf{O}_o(1,n-1)$$

of dimension n and let $f \in C_c^{\infty}(X)$ and $y \in X$. If $z = \mathrm{Exp}_y Y$ $(Y \in D_y^{-})$ we put $r_{yz} = g(Y,Y)^{\frac{1}{2}}$ and consider the integral

$$(13) \qquad (I_-^{\lambda} f)(y) = \frac{1}{H_n(\lambda)} \int_{\bar{D}_y} f(z) \sinh^{\lambda-n}(r_{yz})\, dz\,,$$

where dz is the volume element on X, and

$$(14) \qquad H_n(\lambda) = \pi^{\frac{1}{2}(n-2)} 2^{\lambda-1}\, \Gamma\left(\frac{\lambda}{2}\right) \Gamma\left(\frac{\lambda+2-n}{2}\right).$$

The integral converges for $\mathrm{Re}\,\lambda \geq n$. We transfer the integral in (13) over to D_y via the diffeomorphism $\mathrm{Exp}\,(=\mathrm{Exp}_y)$. Since

$$dz = dr\,d\bar{\omega}^r = dr \left(\frac{\sinh r}{r}\right)^{n-1} d\,\omega^r$$

and since $dr\,d\omega^r$ equals the volume element dZ on D_y we obtain

$$(I_-^{\lambda} f)(y) = \frac{1}{H_n(\lambda)} \int_{D_y} (f \circ \mathrm{Exp})(Z) \left(\frac{\sinh r}{r}\right)^{\lambda-1} r^{\lambda-n} dZ,$$

where $r = g(Z,Z)^{\frac{1}{2}}$. This has the form

$$\frac{1}{H_n(\lambda)} \int_{D_y} h(Z,\lambda)\, r^{\lambda-n}\, dZ, \tag{15}$$

where $h(Z, \lambda)$, as well as each of its partial derivatives with respect to the first argument, is holomophic in λ and h has compact support in the first variable. The methods of Riesz [1949], Ch. III, can be applied to such integrals (15). In particular we find that the function $\lambda \to (I_-^\lambda f)(y)$ which by its definition is holomorphic for $\operatorname{Re}\lambda > n$ admits a holomorphic continuation to the entire λ-plane and that its value at $\lambda = 0$ is $h(0,0) = f(y)$. (In Riesz'' treatment $h(Z,\lambda)$ is independent of λ, but his method still applies). Denoting the holomorphic continuation of (13) by $(I_-^\lambda f)(y)$ we have thus obtained

$$I_-^0 f = f. \tag{16}$$

We would now like to differentiate (13) with respect to y. For this we write the integral in the form $\int_F f(z)\, K(y,z)\, dz$ over a bounded region F which properly contains the intersection of the support of f with the closure of $\overline{D}_y$. The kernel $K(y,z)$ is defined as $\sinh^{\lambda-n} r_{yz}$ if $z \in \overline{D}_y$, otherwise 0. For $\operatorname{Re}\lambda$ sufficiently large, $K(y,z)$ is twice continuously differentiable in y so we can deduce for such λ that $I_-^\lambda f$ is of class C^2 and that

$$(\square I_-^\lambda f)(y) = \frac{1}{H_n(\lambda)} \int_{\overline{D}_y} f(z)\, \square_y (\sinh^{\lambda-n} r_{yz})\, dz\,. \tag{17}$$

Moreover, given $m \in \mathbf{Z}^+$ we can find k such that $I_-^\lambda f \in C^m$ for $\operatorname{Re}\lambda > k$ (and all f). Using Lemma 2.10 and the relation

$$\frac{1}{A(r)}\, \frac{dA}{dr} = (n-1) \coth r$$

we find

$$\square_y(\sinh^{\lambda-n} r_{yz}) = \square_z(\sinh^{\lambda-n} r_{yz})$$

$$= (\lambda-n)(\lambda-1)\sinh^{\lambda-n} r_{yz} + (\lambda-n)(\lambda-2)\sinh^{\lambda-n-2} r_{yz}.$$

We also have

$$H_n(\lambda) = (\lambda-2)(\lambda-n)H_n(\lambda-2)$$

so substituting into (17) we get

$$\square I_-^{\lambda} f = (\lambda-n)(\lambda-1)I_-^{\lambda} f + I_-^{\lambda-2} f.$$

Still assuming $\operatorname{Re}\lambda$ large we can use Green's formula to express the integral

$$(18) \qquad \int_{\overline{D}_y} \left[f(z)\, \square_z(\sinh^{\lambda-n} r_{yz}) - \sinh^{\lambda-n} r_{yz} (\square f)(z) \right] dz$$

as a surface integral over a part of $\overline{C}_y$ (on which $\sinh^{\lambda-n} r_{yz}$ and its first order derivatives vanish) together with an integral over a surface inside $\overline{D}_y$ (on which f and its derivativies vanish). Hence the expression (18) vanishes so we have proved the relations

$$(19) \qquad \square(I_-^{\lambda} f) = I_-^{\lambda}(\square f)$$

$$(20) \qquad I_-^{\lambda}(\square f) = (\lambda-n)(\lambda-1)I_-^{\lambda} f + I_-^{\lambda-2} f$$

for $\operatorname{Re}\lambda > k$, k being some number (independent of f).

Since both sides of (20) are holomorphic in λ this relation holds for all $\lambda \in \mathbb{C}$. We shall now deduce that for each $\lambda \in \mathbb{C}$, $I_-^{\lambda} f \in C^{\infty}(X)$ and (19) holds. For this we observe by iterating (20) that for each $p \in \mathbb{Z}^+$

$$(21) \qquad I_-^{\lambda} f = I_-^{\lambda+2p}(Q_p(\square)f),$$

Q_p being a certain p^{th}-degree polynomial. Choosing p arbitrarily large we deduce from the remark following (17) that $I_-^{\lambda} f \in C^{\infty}(X)$; secondly (19) implies for $\mathrm{Re}\,\lambda + 2p > k$ that

$$\square\, I_-^{\lambda+2p}(Q_p(\square)f) = I_-^{\lambda+2p}(Q_p(\square)\square f).$$

Using (21) again this means that (19) holds for all λ.

Putting $\lambda = 0$ in (20) we get

$$(22) \qquad I_-^{-2} f = \square f - nf.$$

Remark.. In Riesz' paper [1949], p.190, an analog I^{α} of the potentials in Ch.I, § 8, is defined for any analytic Lorentzian manifold. These potentials I^{α} are however different from our I_-^{λ} and satisfy the equation $I^{-2} f = \square f$ in contrast to (22).

We consider next the case

$$X = Q_{+1} = G^+/H = \mathbf{O}_o(2,n-1)/\mathbf{O}_o(1,n-1)$$

and define for $f \in C_c^{\infty}(X)$

$$(23) \qquad (I_+^{\lambda} f)(y) = \frac{1}{H_n(\lambda)} \int_{\overline{D}_y} f(z) \sin^{\lambda-n}(r_{yz})\, dz\,.$$

Again $H_n(\lambda)$ is given by (14) and dz is the volume element. In order to bypass the difficulties caused by the fact that the function $z \to \sin r_{yz}$ vanishes on $\overline{S}^{\pi}(y)$ we assume that f has support disjoint from $\overline{S}^{\pi}(o)$. Then the support of f is disjoint from $\overline{S}^{\pi}(y)$ for all y in some neighborhood of o in X. We can then prove just as before that

(24) $$(I_+^0 f)(y) = f(y)$$

(25) $$(\square I_+^\lambda f)(y) = (I_+^\lambda \square f)(y)$$

(26) $$(I_+^\lambda \square f)(y) = -(\lambda-n)(\lambda-1)(I_+^\lambda f)(y) + (I_+^{\lambda-2} f)(y)$$

for all $\lambda \in \mathbb{C}$. In particular,

(27) $$I_+^{-2} f = \square f + nf.$$

Finally we consider the flat case

$$X = \mathbb{R}^n = G^o/H = \mathbb{R}^n \cdot \mathbf{O}_o(1,n-1)/\mathbf{O}_o(1,n-1)$$

and define

$$(I_o^\lambda f)(y) = \frac{1}{H_n(\lambda)} \int_{\bar{D}_y} f(z)\, r_{yz}^{\lambda-n}\, dz.$$

These are the potentials defined by Riesz in [1949], p.31, who proved

(28) $$I_o^0 f = f, \quad \square I_o^\lambda f = I_o^\lambda \square f = I_o^{\lambda-2} f.$$

§4. Determination of a Function from its Integrals over Lorentzian Spheres.

In a Riemannian manifold a function is determined in terms of its spherical mean values by the simple relation $f = \lim_{r \to 0} M^r f$. We shall now solve the analogous problem for an even-dimensional isotropic Lorentz manifold and express a function f in terms of its orbital integrals $M^r f$. Since the sphere $\bar{S}^r(y)$ do not shrink to a point as $r \to 0$ the formula (cf. Theorem 4.1) below is quite different.

For the solution of the problem we use the geometric description of the wave operator $\square$ developed in §2, particularly its commutation with the orbital integral M^r, and combine this with the results about the generalized Riesz potentials established in §3.

We consider first the negatively curved space $X = G^-/H$. Let $n = \dim X$ and assume n even. Let $f \in C_c^\infty(X)$ and put $F(r) = (M^r f)(y)$. since the volume element dz on $\bar{D}_y$ is given by $dz = dr\, d\bar{\omega}^r$ we obtain from (12) and Lemma 2.9,

$$(I_-^\lambda f)(y) = \frac{1}{H_n(\lambda)} \int_0^\infty \sinh^{\lambda-1} r\, F(r)\, dr\,. \tag{29}$$

Let $Y_1, \ldots, Y_n$ be a basis of X_y such that the Lorentzian structure is given by

$$g_y(Y) = y_1^2 - y_2^2 - \cdots - y_n^2, \qquad Y = \sum_1^n Y_i\,.$$

If $\theta_1, \ldots, \theta_{n-2}$ are geodesic polar coordinates on the unit sphere in $\mathbb{R}^{n-1}$ we put

$$y_1 = -r\cosh\zeta \qquad (0 \leq \zeta < \infty\,, \; 0 < r < \infty)$$

$$y_2 = r\sinh\zeta\,\cos\theta_1$$

$$\vdots$$

$$y_n = r\sinh\zeta\,\sin\theta_1 \ldots \sin\theta_{n-2}\,.$$

Then $(r, \zeta, \theta_1, \ldots, \theta_{n-2})$ are coordinates on the retrograde cone D_y and the volume element on $S^r(y)$ is given by

$$d\omega^r = r^{n-1}\sinh^{n-2}\zeta\; d\zeta\; d\omega^{n-2}$$

where $d\omega^{n-2}$ is the volume element on the unit sphere in $\mathbb{R}^{n-1}$. It follows that

$$d\bar{\omega}^r = \sinh^{n-1} r\,\sinh^{n-2}\zeta\; d\zeta\; d\omega^{n-2}$$

and therefore

$$(30) \quad F(r) = \iint (f \circ \mathrm{Exp})(r,\zeta,\theta_1, \ldots, \theta_{n-2}) \sinh^{n-2}\zeta \, d\zeta \, d\omega_{n-2}$$

where

$$(r,\zeta,\theta_1, \ldots, \theta_{n-2})$$

$$= (-r\cosh\zeta, r\sinh\zeta\cos\theta_1, \ldots, r\sinh\zeta\sin\theta_1 \ldots \sin\theta_{n-2}).$$

Now select A such that $f \circ \mathrm{Exp}$ vanishes outside the sphere $y_1^2 + \cdots + y_n^2 = A^2$ in X_y. Then, in the integral (30) the range of ζ is contained in the interval $(0, \zeta_0)$ where $r^2\cosh^2\zeta_0 + r^2\sinh^2\zeta_0 = A^2$. We see by the substitution $t = r\sinh\zeta$ that the integral expression (30) behaves for small r like

$$\int_0^k \phi(t) \left(\frac{t}{r}\right)^{n-2} (r^2+t^2)^{-\frac{1}{2}} dt$$

where φ is bounded. Therefore if $n>2$ the limit

$$(31) \qquad a = \lim_{r\to 0} \sinh^{n-2} r \, F(r) \qquad (n>2)$$

exists. Similarly, we find for $n=2$ that the limit

$$(32) \qquad b = \lim_{r\to 0} (\sinh r) \, F'(r) \qquad (n=2)$$

exists.

Consider now the case $n>2$. We can rewrite (29) in the form

$$(I_-^\lambda f)(y) = \frac{1}{H_n(\lambda)} \int_0^A \sinh^{n-2} r \, F(r) \sinh^{\lambda-n+1} r \, dr,$$

where $F(A) = 0$. We now evaluate both sides for $\lambda = n-2$. Since $H_n(\lambda)$ has a simple pole for $\lambda = n-2$ the integral has at most a simple pole there and the residue is

$$\lim_{\lambda \to n-2} (\lambda - n+2) \int_0^A \sinh^{n-2} r\, F(r) \sinh^{\lambda-n+1} r\, dr .$$

Here we can take λ real and greater than $n-2$. This is convenient since by (31) the integral is then absolutely convergent and we do not have to think of it as an implicitly given holomorphic extension. We split the integral in two parts

$$(\lambda - n+2) \int_0^A (\sinh^{n-2} r \; F(r) - a) \sinh^{\lambda-n+1} r\, dr$$

$$+ \, a(\lambda - n+2) \int_0^A \sinh^{\lambda-n+1} r\, dr .$$

For the last term we use the relation

$$\lim_{\mu \to 0+} \mu \int_0^A \sinh^{\mu-1} r\, dr = \lim_{\mu \to 0+} \mu \int_0^{\sinh A} t^{\mu-1} (1+t^2)^{-\frac{1}{2}} \, dt = 1$$

by (75) in Ch. I. For the first term we can for each $\varepsilon > 0$ find a $\delta > 0$ such that

$$|\sinh^{n-2} r\, F(r) - a| < \varepsilon \qquad \text{for } 0 < r < \delta .$$

If $N = \max |\sinh^{n-2} r\, F(r)|$ we have for $n-2 < \lambda < n-1$ the estimates

$$\left| (\lambda - n+2) \int_\delta^A (\sinh^{n-2} r\, F(r) - a) \sinh^{\lambda-n+1} r\, dr \right|$$

$$\leq (\lambda - n+2)(N + |a|)(A - \delta) (\sinh \delta)^{\lambda-n+1}$$

$$\left| (\lambda - n+2) \int_0^\delta (\sinh^{n-2} r\, F(r) - a) \sinh^{\lambda-n+1} r\, dr \right|$$

$$\leq \varepsilon (\lambda - n+2) \int_0^\delta r^{\lambda-n+1} \, dr = \varepsilon\, \delta^{\lambda-n+2} .$$

Taking $\lambda - (n-2)$ small enough the right hand side of each of these inequalities is $< 2\varepsilon$. We have therefore proved

$$\lim_{\lambda\to n-2} (\lambda - n+2) \int_0^\infty \sinh^{\lambda-1} r\, F(r)\, dr = \lim_{r\to 0} \sinh^{n-2} r\, F(r).$$

Taking into account the formula for $H_n(\lambda)$ we have proved for the integral (29):

$$(33) \qquad I_-^{n-2} f = (4\pi)^{\frac{1}{2}(2-n)} \frac{1}{\Gamma(\frac{1}{2}(n-2))} \lim_{r\to 0} \sinh^{n-2} r\; M^r f$$

On the other hand, using formula (20) recursively we obtain for $u \in C_c^\infty(X)$

$$I_-^{n-2}(Q(\square)u) = u$$

where

$$Q(\square) = (\square + (n-3)2)(\square + (n-5)4) \cdots (\square + 1(n-2)).$$

We combine this with (33) and use the commutativity $\square M^r = M^r \square$. This gives

$$(34) \qquad u = (4\pi)^{\frac{1}{2}(2-n)} \frac{1}{\Gamma(\frac{1}{2}(n-2))} \lim_{r\to 0} \sinh^{n-2} r\, Q(\square)\, M^r u.$$

Here we can replace $\sinh r$ by r and can replace $\square$ by the operator

$$\square_r = \frac{d^2}{dr^2} + (n-1)\coth r \frac{d}{dr}$$

because of Lemma 2.10 and Cor. 2.14.

For the case $n=2$ we have by (29)

$$(35) \qquad (I_-^2 f)(y) = \frac{1}{H_2(2)} \int_0^\infty \sinh r\, F(r)\, dr .$$

This integral which in effect only goes from 0 to A is absolutely convergent because our estimate of (30) shows (for $n=2$) that $rF(r)$ is bounded near $r=0$. But using (20), (35), Lemma 2.10, Theorem 2.11, and Cor. 2.14, we obtain for $u \in C_c^\infty(X)$,

$$u = I_-^2 \square u = \tfrac{1}{2} \int_0^\infty \sinh r\, M^r \square u\, dr$$

$$= \tfrac{1}{2} \int_0^\infty \sinh r\, \square M^r u\, dr = \tfrac{1}{2} \int_0^\infty \sinh r \left(\frac{d^2}{dr^2} + \coth r \frac{d}{dr} \right) M^r u\, dr$$

$$= \tfrac{1}{2} \int_0^\infty \frac{d}{dr} \left(\sinh r \frac{d}{dr} M^r u \right) dr = -\tfrac{1}{2} \lim_{r \to 0} \sinh r \frac{d(M^r u)}{dr} .$$

This is the substitute for (34) in the case $n = 2$.

The spaces G^+/H and G^0/H can be treated in the same manner. We have thus proved the following principal result of this chapter.

THEOREM 4.1. *Let X be one of the isotropic Lorentzian manifolds G^-/H, G^0/H, G^+/H. Let κ denote the curvature of X ($\kappa = -1, 0, +1$) and assume $n = \dim X$ to be even. Put*

$$Q(\square) = (\square - \kappa(n-3)2)(\square - \kappa(n-5)4) \cdot \cdot \cdot (\square - \kappa 1(n-2)).$$

Then if $u \in C_c^\infty(X)$,

$$u = (4\pi)^{\frac{1}{2}(2-n)} \frac{1}{\Gamma(\frac{1}{2}(n-2))} \lim_{r \to 0} r^{n-2} Q(\square_r)(M^r u), \qquad (n \neq 2)$$

$$u = -\tfrac{1}{2} \lim_{r \to 0} r \frac{d}{dr} (M^r u) \qquad (n = 2)$$

Here $\square$ is the Laplace-Beltrami operator and $\square_r$ its radial part

$$\square_r = \frac{d^2}{dr^2} + \frac{1}{A(r)} \frac{dA}{dr} \frac{d}{dr} .$$

BIBLIOGRAPHICAL NOTES

§1. The construction of the constant curvature spaces (Theorem 1.3 and 1.5) was given by the author ([1959], [1961]). The proof of Lemma 1.4 on the connectivety is adapted from Boerner [1955]. For more information on isotropic manifolds (there is more than one definition) see Tits [1955], p.183 and Wolf [1967].

§§ 2-4. This material is based on Ch. IV in Helgason [1959]. It is a problem of considerable interest to extend Theorem 4.1 to affine symmetric spaces G/H (G semisimple). As explained in the discussion of Problem D in Ch. II, §3, this would incorporate the representation of a function on G by its integral over the conjugacy classes and their translates.

REFERENCES

Amemiya, I. and T. Ando [1965], Convergence of random products of contractions in Hilbert space. Acta Sci. Math. (Szeged), 26 (1965), 239-244.

Araki, S. I. [1962], On root systems and an infinitesimal classification of irreducible symmetric spaces. J. Math. Osaka City Univ., 13 (1962), 1-34.

Besse, A. [1978], "Manifolds all of whose geodesics are closed." Ergebnisse der Math., Vol. 93, Springer, New York, 1978.

Boerner, H. [1955], "Darstellungen von Gruppen." Springer Verlag, Heidelberg, 1955.

Borovikov, W. A. [1959], Fundamental solutions of linear partial differential equations with constant coefficients. Trudy Moscov Mat. Obšč., 8 (1959), 199-257.

Bracewell, R. N. and A. C. Riddle [1967], Inversion of fan beam scans in radio astronomy, Astro Phys. J., 150 (1967), 427-434.

Cartan, É. [1927], Sur certaines formes riemanniennes remarquables des géometries a groupe fondamental simple. Ann. Sci. École Norm. Sup., 44 (1927), 345-467.

Chern, S. S. [1942], On integral geometry in Klein spaces. Ann. of Math. 43 (1942), 178-189.

Chevalley, C. [1946], "Theory of Lie Groups," Vol. I, Princeton Univ. Press, Princeton, New Jersey, 1946.

Cormack, A, M. [1963], [1964], Representation of a function by its line integrals, with some radiological application I, II. Journal of Applied Physics 34 (1963), 2722-2727; 35 (1964), 2908-2912.

Courant, R. and A. Lax [1955], Remarks on Cauchy's problem for hyperbolic partial differential equations with constant coefficients in several independent variables. Comm. Pure Appl. Math. 8 (1955), 497-502.

Coxeter, H. S. M. [1957], "Non-Euclidean Geometry," Univ. of Toronto Press, Toronto, 1957.

Flensted-Jensen, M. [1977], Spherical functions on a simply connected semisimple Lie group, II. Math. Ann. 228 (1977), 65-92.

Fuglede, B. [1958], An integral formula. Math. Scand. 6 (1958), 207-212.

Funk, P. [1916], Über eine geometrische Anwendung der Abelschen Integralgleichung. Math. Ann. 77 (1916), 129-135.

Gårding, L. [1961], Transformation de Fourier des distributions homogènes. Bull. Soc. Math. France, 89 (1961), 381-428.

Gelfand, I. M. and M. I. Graev [1955], Analogue of the Plancherel formula for the classical groups. Trudy Moscov. Mat. Obšč. 4 (1955), 375-404.
[1968], Complexes of straight lines in the space $\mathbb{C}^n$. Functional analysis and its applications. 2 (1968), 39-52.

Gelfand, I. M., M. I. Graev and N. Vilenkin, [1962], "Generalized Functions, Vol, 5", Engl. Transl. Academic Press, 1966.

Gelfand, I. M. and G. E. Schilov [1959], "Verallgemeinerte Funktionen" Vol. I. German Transl. VEB, Berlin 1960.

Gelfand, I. M. and S. J. Shapiro [1955], Homogeneous functions and their applications. Uspehi. Mat. Nauk. 10 (1955), 3-70.
[1969], Differential forms and integral geometry. Functional Anal. Appl. 3(1969), 24-40.

Godement, R. [1966], The decomposition of $L^2(G/\Gamma)$ for $\Gamma = SL(2, \mathbb{Z})$, Proc. Sympos. Pure Math. Vol. 9, Amer. Math. Soc. 1966, 211-224.

Guillemin, V. [1976], Radon transform on Zoll surfaces. Advan. Math. 22 (1976), 85-119.

Halperin, I. [1962], The product of projection operators. Acta Sci. Math. (Szeged) 23 (1962), 96-99.

Hamaker, C. and D. C. Solmon [1978], The angles between the null spaces of X-rays. J, Math. Anal. and Appl. 62 (1978), 1-23.

Harish-Chandra [1957], A formula for semisimple Lie groups. Amer. J. Math. 79 (1957), 733-760.

Helgason, S. [1959], Differential operators on homogeneous spaces. Acta Math. 102 (1959), 239-299.

[1961], Some remarks on the exponential mapping for an affine connection. Math. Scand. 9 (1961), 129-146.

[1963], Duality and Radon transform for symmetric spaces. Amer. J. Math. 85 (1963), 667-692.

[1964 A], A duality in integral geometry; some generalizations of the Radon transform. Bull, Amer. Math. Soc. 70 (1964), 435-446.

[1964 B], Fundamental solutions of invariant differential operators on symmetric spaces. Amer. J. Math. 86 (1964), 565-601.

[1965 A], The Radon transform on Euclidean spaces, compact two-point homogeneous spaces and Grassmann manifolds. Acta Math. 113 (1965), 153-180.

[1965 B], A duality in integral geometry on symmetric spaces. Proc. U.S. - Japan Seminar in Differential Geometry, Kyoto 1965. Nippon Hyoransha, Tokyo 1966, 37-56.

[1973], The surjectivity of invariant differential operators on symmetric spaces I. Ann of Math. 98 (1973), 451-479.

[1978], "Differential Geometry, Lie Groups and Symmetric Spaces." Academic Press, New York, 1978.

[1980 A], Support of Radon transforms. Advan. Math.

(to appear).
[1980 B], The X-ray transform on a symmetric space. Proc. Conf. Diff. Geom. and Global Analysis, Berlin 1979, Lecture Notes. Springer, New York, 1980.

Herglotz, G. [1931], Notes of Lectures on "Mechanik der Kontinua," Göttingen, 1931.

Hörmander, L. [1963], "Linear Differential Operators." Springer Verlag, New York, 1963.

John, F. [1934], Bestimmung einer Funktion aus ihren Integralen über gewisse Mannigfaltigkeiten. Math. Ann. 109 (1934), 488-520.
[1935], Abhängigkeit zwischen den Flächenintegralen einer stetigen Funktion. Math. Ann. 111 (1935), 541-559.
[1938], The ultrahyperbolic differential equation with 4 independent variables. Duke Math. J. 4 (1938), 300-322.
[1955], "Plane Waves and Spherical Means applied to Partial differential equations." Interscience, New York, 1955.

Lax, P. D. and R. S. Phillips [1967], "Scattering Theory." Academic Press, New York, 1967.
[1979], Translation representations for the solution of the non-Euclidean wave equation. Comm. Pure Appl. Math. 32 (1979), 617-667.

Lichnèrowicz, A. and A. G. Walker [1945], Sur les espaces Riemanniens harmoniques de type hyperbolique normal. C. R. Acad. Sci. Paris 221 (1945), 397-396.

Ludwig, D. [1966], The Radon transform on Euclidean space. Comm. Pure Appl. Math. 23 (1966), 49-81.

Matsumoto, H. [1971], Quelques remarques sur les espaces riemanniens isotropes. C. R. Acad. Sci. Paris 272 (1971), 316-319.

Michel, L. [1972], Sur certains tenseurs symétriques des projectifs

réels. J. Math. pures et appl. 51 (1972), 275-293.
[1973], Problèmes d'analyse geométrique liés a la conjecture de Blaschke. Bull. Soc. Math. France 101 (1973), 17-69.

Nagano, T. [1959], Homogeneous sphere bundles and the isotropic Riemannian manifolds. Nagoya Math. J. 15 (1959), 29-55.

Penrose, R. [1967], Twistor algebra. J. Math. Phys. 8 (1967), 345-366.

Quinto, E.T. [1980], The dependence of the generalized Radon transform on defining measures. Trans. Amer. Math. Soc. 257 (1980), 331-346.

Radon, J. [1917], Über die Bestimmung von Funktionen durch ihre Integralwerte längs gewisser Mannigfaltigkeiten. Ber. Verh. Sächs. Akad. Wiss. Leipzig, Math-Nat. kl. 69 (1917), 262-277.

Riesz, M. [1949], L'intégrale de Riemann-Liouville et le problème de Cauchy. Acta Math. 81 (1949), 1-223.

Santalo, L. [1976], "Integral Geometry and Geometric Probability," Addison Wesley, Reading, 1976.

Schwartz, L. [1966], "Théorie des Distributions" Hermann, Paris, 1966.

Selberg, A. [1962], Discontinuous groups and harmonic analysis. Proc. Internat. Congr. Math., Stockholm, (1962), 177-189.

Semyanistyi, V. I. [1960], On some integral transforms in Euclidean space. Soviet Math. Dokl. 1 (1960), 1114-1117.
[1961], Homogeneous functions and some problems of integral geometry in spaces of constant curvature. Soviet Math. Dokl. 2 (1961), 59-62.

Shepp, L. A. and J. B. Kruskal [1978], Computerized tomography; the new medical X-ray technology. Amer. Math. Monthly (1978), 420-438.

Smith, K. T. and D. C. Solmon [1975], Lower-dimensional integrability of L^2 functions. J. Math. Anal. Appl. 51 (1975), 539-549.

Smith, K. T., D. C. Solmon and S. L. Wagner [1977], Practical and mathematical aspects of the problem of reconstructing objects from radiographs. Bull. Amer. Math. Soc. 83 (1977), 1227-1270.

Solmon, D. C. [1976], The X-ray transform, J. Math. Anal. Appl. 56 (1976), 61-83.

Tits, J. [1955], Sur certains classes d'espaces homogènes de groupes de Lie. Acad. Roy. Belg. Cl. Sci. Mem. Coll. 29 (1955), No. 3.

Trèves, F. [1963], Équations aux dérivées partielles inhomogènes a coefficients constants dépendent de paramètres. Ann. Inst. Fourier, Grenoble 13 (1963), 123-138.
[1967], "Topological Vector Spaces, Distributions and Kernels." Academic Press, New York, 1967.

Wang, H. C. [1952], Two-point homogeneous spaces. Ann. of Math. 55 (1952), 177-191.

Weiss, B. [1967], Measures that vanish on half spaces. Proc. Amer. Math. Soc. (1967), 123-126.

Wells, Jr., R. [1979], Complex manifolds and mathematical physics. Bull. Amer. Math. Soc. 1 (1979), 275-442.

Whittaker, E. T. and G. N. Watson [1927], "A Course of Modern Analysis." Cambr. Univ. Press, 1927.

Wolf, J. A. [1967], "Spaces of Constant Curvature." McGraw-Hill, New York, 1967.

NOTATIONAL CONVENTIONS

Algebra. As usual, $\mathbb{R}$ and $\mathbb{C}$ denote the fields of real and complex numbers, respectively, and $\mathbb{Z}$ the ring of integers. Let

$$\mathbb{R}^+ = \{t \in \mathbb{R} : t \geq 0\}, \quad \mathbb{Z}^+ = \mathbb{Z} \cap \mathbb{R}^+.$$

If $\alpha \in \mathbb{C}$, $\text{Re}\alpha$ denotes the real part of α, $|\alpha|$ its modulus.

If G is a group, $A \subset G$ a subset and $g \in G$ an element, we put

$$A^g = \{gag^{-1} : a \in A\}, \quad g^A = \{aga^{-1} : a \in A\}.$$

The group of real matrices leaving invariant the quadratic form

$$x_1^2 + \ldots + x_p^2 - x_{p+1}^2 - \ldots - x_{p+q}^2$$

is denoted by $\mathbf{O}(p,q)$. We put $\mathbf{O}(n) = \mathbf{O}(o,n) = \mathbf{O}(n,o)$, and write $\mathbf{U}(n)$ for the group of $n \times n$ unitary matrices. The group of isometries of Euclidean n-space $\mathbb{R}^2$ is denoted by $M(n)$.

Geometry. The $(n-1)$-dimensional unit sphere in $\mathbb{R}^n$ is denoted by $\mathbb{S}^{n-1}$, Ω_n denotes its area. The n-dimensional manifold of hyperplanes in $\mathbb{R}^n$ is denoted by $\mathbb{P}^n$. If $0 < d < n$ the manifold of d-dimensional planes in $\mathbb{R}^n$ is denoted by $G(d,n)$; we put $G_{d,n} = \{\sigma \in G(d,n) : o \in \sigma\}$. In a metric space, $B^r(x)$ denotes the open ball with center x and radius r; $S^r(x)$ denotes the corresponding sphere. For $\mathbb{P}^n$ we use the notation $\beta^A(0)$ for the set of hyperplanes $\xi \subset \mathbb{R}^n$ of distance $<A$ from 0.

Analysis. If X is a topological space, $C(X)$ (resp. $C_c(X)$) denotes the space of complex-valued continuous functions (resp. of compact support). If X is a manifold, we denote:

$$C^m(X) = \begin{Bmatrix}\text{complex-valued m-times continuously}\\ \text{differentiable functions on } X\end{Bmatrix}$$

$$C^\infty(X) = \mathcal{E}(X) = \bigcap_{m>0} C^m(X).$$

$$C_c^\infty(X) = \mathcal{D}(X) = C_c(X) \cap C^\infty(X).$$

$\mathcal{D}'(X)$ = {distributions on X}.

$\mathcal{E}'(X)$ = {distributions on X of compact support}.

$\mathcal{D}_A(X) = \{f \in \mathcal{D}(X) : \text{support } f \subset A\}$.

$\mathcal{S}(\mathbb{R}^n)$ = {rapidly decreasing functions on $\mathbb{R}^n$}

$\mathcal{S}'(\mathbb{R}^n)$ = {tempered distributions on $\mathbb{R}^n$}.

The subspaces $\mathcal{D}_H, \mathcal{S}_H, \mathcal{S}^*, \mathcal{S}_o$ of $\mathcal{S}$ are defined page 6, 11 and 12.

While the functions considered are usually assumed to be complex-valued, we occasionally use the notation above for spaces of real-valued functions.

The Radon transform and its dual are denoted by $f \longrightarrow \hat{f}$, $\phi \longrightarrow \check{\phi}$, the Fourier transform by $f \longrightarrow \tilde{f}$ and the Hilbert transform by $\mathcal{H}$.

I^α, I_-^λ, I_o^λ and I_+^λ denote Riesz potentials and their generalizations, M^r the mean value operator and orbital integral, L the Laplacian on $\mathbb{R}^n$ and the Laplace-Beltrami operator on a pseudo-Riemannian manifold. The operators $\square$ and Λ operate on certain function spaces on $\mathbb{P}^n$ (page 3 and 25); $\square$ is also used for the Laplace-Beltrami operator on a Lorentzian manifold.

SUBJECT INDEX

APPENDIX

Reprinted from:
Ber. Verh. Sächs. Akad. Wiss. Leipzig,
Math-Nat. kl. 69 (1917), 262-277.

SITZUNG VOM 30. APRIL 1917.

Über die Bestimmung von Funktionen durch ihre Integralwerte längs gewisser Mannigfaltigkeiten.

Von

JOHANN RADON.

Integriert man eine geeigneten Regularitätsbedingungen unterworfene Funktion zweier Veränderlichen x, y — eine *Punktfunktion* $f(P)$ in der Ebene — längs einer beliebigen Geraden g, so erhält man in den Integralwerten $F(g)$ eine *Geradenfunktion*. Das in Abschnitt A vorliegender Abhandlung gelöste Problem ist die Umkehrung dieser linearen Funktionaltransformation, d. h. es werden folgende Fragen beantwortet: kann jede, geeigneten Regularitätsbedingungen genügende Geradenfunktion auf diese Weise entstanden gedacht werden? Wenn ja, ist dann f durch F eindeutig bestimmt und wie kann es ermittelt werden?

Im Abschnitte B gelangt das dazu in gewisser Hinsicht duale Problem der Bestimmung einer Geradenfunktion $F(g)$ aus ihren Punktmittelwerten $f(P)$ zur Lösung.

Schließlich werden im Abschnitte C gewisse Verallgemeinerungen kurz besprochen, zu denen insbesondere die Betrachtung nichteuklidischer Mannigfaltigkeiten sowie höherer Räume Anlaß gibt.

Die Behandlung dieser an sich interessanten Probleme gewinnt ein erhöhtes Interesse durch die zahlreichen Beziehungen, die zwischen diesem Gegenstande und der Theorie des logarithmischen und NEWTONschen Potentiales bestehen, auf die an den bezüglichen Stellen zu verweisen sein wird.

A. Bestimmung einer Punktfunktion in der Ebene aus ihren geradlinigen Integralwerten.

1. Es sei $f(x, y)$ eine für alle reellen Punkte $P = [x, y]$ erklärte reelle Funktion, die folgende Regularitätsbedingungen erfülle:

a_1) $f(x, y)$ sei stetig.

b_1) Es konvergiere das über die ganze Ebene zu erstreckende Doppelintegral

$$\iint \frac{|f(x, y)|}{\sqrt{x^2 + y^2}} dx\, dy.$$

c_1) Wird für einen beliebigen Punkt $P = [x, y]$ und jedes $r \geqq 0$

$$\bar{f}_P(r) = \frac{1}{2\pi} \int_0^{2\pi} f(x + r\cos\varphi,\ y + r\sin\varphi)\, d\varphi$$

gesetzt, so gelte für jeden Punkt P:

$$\lim_{r \to \infty} \bar{f}_P(r) = 0.$$

Dann gelten folgende Sätze:

Satz I: Der geradlinige Integralwert von f für die Gerade g mit der Gleichung $x\cos\varphi + y\sin\varphi = p$, der durch

$$\text{(I)}\quad F(p, \varphi) = F(-p, \varphi + \pi) = \int_{-\infty}^{+\infty} f(p\cos\varphi - s\sin\varphi,\ p\sin\varphi + s\cos\varphi)\, ds$$

gegeben ist, ist „im allgemeinen“ vorhanden; das soll heißen: auf jedem Kreise bilden die Berührungspunkte jener Tangenten, für welche F nicht existiert, eine Menge vom linearen Maße Null:

Satz II: Bildet man den Mittelwert von $F(p, \varphi)$ für die Tangenten des Kreises mit dem Zentrum $P = [x, y]$ und dem Radius q:

$$\text{(II)}\qquad F_P(q) = \frac{1}{2\pi} \int_0^{2\pi} F(x\cos\varphi + y\sin\varphi + q,\ \varphi)\, d\varphi,$$

so konvergiert dieses Integral für alle P, q absolut.

Satz III: Der Wert von f ist durch F eindeutig bestimmt und läßt sich folgendermaßen berechnen:

$$\text{(III)}\qquad f(P) = -\frac{1}{\pi} \int_0^{\infty} \frac{dF_P(q)}{q}.$$

Dabei ist das Integral im STIELTJES schen Sinne zu verstehen und kann auch durch die Formel:

$$\text{(III')}\qquad f(P) = \frac{1}{\pi}\lim_{\varepsilon\to 0}\left(\frac{F_p(\varepsilon)}{\varepsilon} - \int_{\varepsilon}^{\infty}\frac{F_p(q)}{q^2}\,dq\right)$$

definiert werden.

Indem wir zum Beweise schreiten, bemerken wir vorweg, daß die Bedingungen $a_1 - c_1$ gegenüber Bewegungen der Ebene invariant sind. Wir können also den Punkt [0, 0] immer als Repräsentanten irgendeines Punktes der Ebene betrachten.

Man erkennt nun das Doppelintegral:

$$(1)\qquad \iint\limits_{x^2+y^2>q^2}\frac{f(x,y)}{\sqrt{x^2+y^2-q^2}}\,dx\,dy$$

als absolut konvergent. Vermöge der Transformation

$$x = q\cos\varphi - s\sin\varphi,\quad y = q\sin\varphi + s\cos\varphi$$

geht dasselbe über in:

$$\int_0^{2\pi}d\varphi\int_0^{\infty}f(q\cos\varphi - s\sin\varphi,\ q\sin\varphi + s\cos\varphi)\,ds$$

$$= \int_0^{2\pi}d\varphi\int_{-\infty}^{0}f(q\cos\varphi - s\sin\varphi,\ q\sin\varphi + s\cos\varphi)\,ds,$$

so daß man seinen Wert auch durch

$$\frac{1}{2}\int_0^{2\pi}d\varphi\int_{-\infty}^{+\infty}f(q\cos\varphi - s\sin\varphi,\ q\sin\varphi + s\cos\varphi)\,ds = \frac{1}{2}\int_0^{2\pi}F(q,\varphi)\,d\varphi$$

ausdrücken kann. Nach bekannten Eigenschaften der absolut konvergenten Doppelintegrale folgen hieraus die Behauptungen der Sätze I und II.

Um auf die Formel (III) zu kommen, kann man folgenden Weg einschlagen: Einführung von Polarkoordinaten in (1) liefert

$$\int^{\infty}dr\int_0^{2\pi}\frac{f(r\cos\varphi,\ r\sin\varphi)}{\sqrt{r^2-q^2}}\,d\varphi$$

oder mit Hilfe der Mittelwertsbezeichnung von $\mathfrak{c}_1$:

$$2\pi\int_q^\infty \frac{\bar f_0(r)\,dr}{\sqrt{r^2-q^2}}.$$

Verglichen mit dem zuletzt erhaltenen Werte von (1) resultiert:

$$\bar F_0(q) = 2\int_q^\infty \frac{\bar f_0(r)\,dr}{\sqrt{r^2-q^2}}. \tag{2}$$

Führt man in dieser Integralgleichung erster Art die Variablen $r^2 = v$, $q^2 = u$ ein, so kann man sie leicht nach Art der bekannten Abelschen lösen und erhält die Formel (III) für

$$\bar f_0(0) = f(0,0).$$

Bei dieser Ableitung erscheint es aber schwer, ohne weitere Bedingungen für f auszukommen, daher geben wir einer direkten Verifikation den Vorzug.

Zunächst ist, um die Gleichheit der Ausdrücke (III) und (III′) zu zeigen, zu beweisen, daß:

$$\lim_{q\to\infty} \frac{\bar F_0(q)}{q} = 0.$$

Vermöge (2) ist

$$\left|\frac{\bar F_0(q)}{q}\right| \leqq \frac{2}{q}\left|\int_q^{2q} \frac{\bar f_0(r)}{\sqrt{r^2-q^2}}\right| + \frac{2}{q}\left|\int_{2q}^\infty \frac{\bar f_0(r)\,r\,dr}{\sqrt{r^2-q^2}}\right|$$

$$\leqq 2\sqrt{3}\,|\bar f_0(t)| + \frac{4}{\sqrt{3}}\int_{2q}^\infty |\bar f_0(r)|\,dr \qquad (q<t<2q)$$

und dies konvergiert wegen $\mathfrak{b}_1$ und $\mathfrak{c}_1$ für $q\to\infty$ gegen Null.

Die rechte Seite von (III′) geht nun durch Einführung von (2) über in:

$$\frac{2}{\pi}\lim_{\varepsilon\to 0}\left[\frac{1}{\varepsilon}\int_\varepsilon^\infty \frac{r\,\bar f_0(r)}{\sqrt{r^2-\varepsilon^2}}\,dr - \int_\varepsilon^\infty \frac{dq}{q^2}\int_q^\infty \frac{r\,\bar f_0(r)}{\sqrt{r^2-q^2}}\,dr\right].$$

Vertauscht man in dem zweiten Integral die Integrationsfolge, so kann man nach q integrieren, erkennt dabei das Integral als absolut konvergentes Doppelintegral, was die Vertauschung rechtfertigt, und findet für obigen Ausdruck den Wert

$$\frac{2}{\pi} \lim_{\varepsilon \to 0} \int_{\varepsilon}^{\infty} \frac{f_0(r)}{r\sqrt{r^2 - \varepsilon^2}} dr,$$

was tatsächlich $f_0(0) = f(0, 0)$ liefert, wie unschwer zu zeigen ist.

2. Es sei $F(p, \varphi) = F(-p, \varphi + \pi)$ eine Geradenfunktion, die folgende Regularitätsbedingungen erfüllt:

a_2) F und die Ableitungen F_p, F_{pp}, F_{ppp}, F_φ, $F_{p\varphi}$, $F_{pp\varphi}$ seien für alle $[p, \varphi]$ stetig.

b_2) F, F_φ, pF_p, $pF_{p\varphi}$ und pF_{pp} konvergieren für $p \to \infty$ gleichmäßig in φ gegen Null.

c_2) Die Integrale:

$$\int_0^\infty F_{pp}\, lp\, dp, \quad \int_0^\infty F_{ppp}\, p\, lp\, dp, \quad \int_0^\infty F_{pp\varphi}\, p\, lp\, dp$$

konvergieren absolut und gleichmäßig in φ.

Dann können wir beweisen:

Satz IV: Bildet man nach (III) bzw. (III') $f(P)$, so erfüllt dieses die Bedingungen a_1, b_1, c_1 und liefert als geradlinige Integralwerte das vorgelegte $F(p, \varphi)$. Infolge Satz III ist es die einzige derartige Funktion.

Führen wir in (III) Polarkoordinaten ein, so ergibt sich:

$$f(\varrho \cos\psi, \varrho \sin\psi) = -\frac{1}{2\pi^2} \int_0^\infty \frac{dp}{p} \int_0^{2\pi} F_p(\varrho \cos\omega + p, \omega + \psi)\, d\omega$$

$$= \frac{1}{2\pi^2} \int_0^\infty lp\, dp \int_0^{2\pi} F_{pp}(p + \varrho \cos\omega, \omega + \psi)\, d\omega.$$

Es ist nämlich:

$$\int_0^{2\pi} F_p(\varrho \cos\omega + p, \omega + \psi)\, d\omega = \int_0^{2\pi} F_p(\varrho \cos\omega, \omega + \psi)\, d\omega$$

$$+ \int_0^{2\pi} d\omega \int_0^p F_{pp}(\varrho \cos\omega + t, \omega + \psi)\, dt$$

und der erste Bestandteil ist wegen $F(p, \varphi) = F(-p, \varphi + \pi)$ gleich Null. Deshalb konvergiert das Produkt des Integrals mit lp

für $p \to 0$ gegen Null. Auf Grund derselben Eigenschaft von F folgt weiter:

$$(3)\quad f(\varrho\cos\psi, \varrho\sin\psi) = \frac{1}{2\pi^2}\int_0^\pi d\omega \cdot \int_{-\infty}^{+\infty} F_{pp}(p, \omega+\psi)\, l\,|p-\varrho\cos\omega|\,dp.$$

Es genügt nun, wenn wir zeigen:

$$(4)\qquad \int_{-\infty}^{+\infty} f(\varrho, 0)\,d\varrho = F\left(0, \frac{\pi}{2}\right),$$

da die Bedingungen a_2—c_2 gegenüber Bewegungen invariant sind. Wir setzen:

$$F(p, \varphi) = F\left(p, \frac{\pi}{2}\right) + \cos\varphi \cdot G(p, \varphi).$$

G genügt leicht angebbaren Regularitätsbedingungen. Es zerfällt nun dieser Zerlegung zufolge $f(\varrho, 0)$ in zwei Bestandteile $f_1(\varrho)$ und $f_2(\varrho)$, die getrennt zu untersuchen sind. Wegen:

$$\int_0^\pi l\,|p-\varrho\cos\omega|\,d\omega = \begin{cases} \pi l\,\dfrac{|p|+\sqrt{p^2-\varrho^2}}{2}, & |p|>|\varrho| \\ \pi l\,\dfrac{|\varrho|}{2}, & |p|\leqq|\varrho| \end{cases}$$

ergibt sich:

$$f_1(\varrho) = \frac{1}{2\pi^2}\int_0^\pi d\omega \int_{-\infty}^{+\infty} F_{pp}\left(p, \frac{\pi}{2}\right) l\,|p-\varrho\cos\omega|\,dp$$

$$= \frac{1}{2\pi}\int_{|\varrho|}^{\infty} F_{pp}\left(p, \frac{\pi}{2}\right) l\,\frac{|p|+\sqrt{p^2-\varrho^2}}{|\varrho|}\,dp$$

$$+ \frac{1}{2\pi}\int_{-\infty}^{-|\varrho|} F_{pp}\left(p, \frac{\pi}{2}\right) l\,\frac{|p|+\sqrt{p^2-\varrho^2}}{|\varrho|}\,dp.$$

Dies ist nun nach ϱ von $-\infty$ bis $+\infty$ absolut integrabel, wie durch Vertauschung der Integrationsfolge ersichtlich wird. Als Wert des Integrales erhält man:

$$\int_{-\infty}^{+\infty} f_1(\varrho)\,d\varrho = \frac{1}{2\pi}\int_{-\infty}^{+\infty} F_{pp}\left(p, \frac{\pi}{2}\right)\int_{-|p|}^{+|p|} l\,\frac{|p|+\sqrt{p^2-\varrho^2}}{|\varrho|}\,d\varrho$$

$$= \frac{1}{2}\int_{-\infty}^{+\infty} F_{pp}\left(p, \frac{\pi}{2}\right)|p|\,dp = F\left(0, \frac{\pi}{2}\right)$$

Was nun $f_2(\varrho)$ anbelangt, so werden wir zeigen, daß es ebenfalls absolut integrabel ist und von $-\infty$ bis $+\infty$ integriert Null ergibt.

Wir können nämlich $f_2(\varrho)$ folgendermaßen schreiben:

$$f_2(\varrho) = \frac{1}{2\pi^2}\int_0^{\pi} d\omega \int_{-\infty}^{+\infty} G_{pp}(p,\omega)\, l\,|p - \varrho\cos\omega| \cdot \cos\omega\, d\omega$$

$$= \frac{1}{2\pi^2}\int_0^{\pi} d\omega \int_{-\infty}^{+\infty} G_{pp}(p,\omega)\left[l\left|\frac{p-\varrho\cos\omega}{\varrho\cos\omega}\right|\cos\omega + \frac{\varrho p\cos^2\omega}{1+\varrho^2\cos^2\omega}\right]dp,$$

da die hinzugefügten Glieder integriert Null ergeben, und in dieser Gestalt führt die Integration nach ϱ auf ein *absolut* konvergentes dreifaches Integral. Es ist nämlich:

$$\int_{-\infty}^{+\infty}\left|\cos\omega\, l\left|\frac{p-\varrho\cos\omega}{\varrho\cos\omega}\right| + \frac{\varrho p\cos^2\omega}{1+\varrho^2\cos^2\omega}\right| d\varrho$$

$$= |p|\int_{-\infty}^{+\infty}\left| l\left|1-\frac{1}{\tau}\right| + \frac{p^2\tau}{1+p^2\tau^2}\right| d\tau = \lambda(p)$$

mit

$$\lim_{p\to\infty}\frac{\lambda(p)}{|p|\,l\,|p|} = 2.$$

Die Integration nach ϱ liefert als Wert des Integrales:

$$\int_{-\infty}^{+\infty} f_2(\varrho)\,d\varrho = 0,$$

womit (4) bewiesen ist.

Wir haben noch zu zeigen, daß f den Forderungen a_1—c_1 genügt.

Die Stetigkeit folgt aus der Darstellung (3) wegen der Voraussetzungen a_2—c_2. Forderung b_1 ist gleichfalls erfüllt, denn

$$\int_{-\infty}^{+\infty} |f(\varrho\cos\psi, \varrho\sin\psi)|\, d\varrho$$

ist, wie man leicht sieht, nach ψ integrierbar.

Um noch c_1 nachzuweisen, bilden wir:

$$f_0(\varrho) = \frac{1}{2\pi}\int_0^{2\pi} f(\varrho\cos\psi,\ \varrho\sin\psi)\,d\varphi$$

$$= \frac{1}{4\pi^3}\int_0^{\pi} d\omega \int_0^{2\pi} d\psi \int_{-\infty}^{+\infty} F_{pp}(p,\psi)\, l\,|p-\varrho\cos\omega|\,dp$$

$$= \frac{1}{4\pi^2}\int_0^{2\pi} d\psi \Bigg[\int_{-\infty}^{-\varrho} F_{pp}(p,\psi)\, l\,\frac{|p|+\sqrt{p^2-\varrho^2}}{2}\,dp$$

$$+\int_{\varrho}^{+\infty} F_{pp}(p,\psi)\, l\,\frac{|p|+\sqrt{p^2-\varrho^2}}{2}\,dp$$

$$+ F_p(\varrho,\psi)\, l\,\frac{\varrho}{2} - F_p(\varrho,\psi)\, l\,\frac{\varrho}{2}\Bigg],$$

woraus man die Richtigkeit von c_1 erkennt. Damit ist Satz IV bewiesen.

B. Bestimmung einer Geradenfunktion aus ihren Punktmittelwerten.

3. $F(p,\varphi) = F(-p,\varphi+\pi)$ sei eine Geradenfunktion, die folgende Regularitätsbedingungen erfülle:

a_3) F, F_φ, F_p seien stetig, $|F_\varphi| < M$ für alle p, φ.

b_3) $F_p \cdot l|p|$ konvergiere für $p \to \infty$ gleichmäßig in φ gegen Null

c_3) $\int_{-\infty}^{+\infty} |F_p| \cdot l|p|\,dp$ konvergiere gleichmäßig in φ.

Diese Bedingungen sind wieder gegenüber Bewegungen invariant.

Wir bilden den Punktmittelwert von $F(p,\varphi)$ für $P=[x,y]$:

$$f(x,y) = \frac{1}{\pi}\int_{-\frac{\pi}{2}}^{+\frac{\pi}{2}} F(x\cos\varphi + y\sin\varphi,\ \varphi)\,d\varphi. \tag{5}$$

Dann gilt:

Satz V: Durch Angabe von f ist F eindeutig bestimmt und zwar ist:

$$(\mathrm{V}) \qquad F\left(0, \frac{\pi}{2}\right) = -\frac{1}{2\pi}\int\limits_{-\infty}^{+\infty}\frac{dx}{x}\int\limits_{-\infty}^{+\infty} f_x(x, y)\,dy,$$

wo das Integral nach x als CAUCHY*scher Hauptwert zu deuten ist und der Wert von F für irgendeine andere Gerade aus der angeschriebenen Formel durch eine entsprechende Bewegung abgeleitet werden kann.*

Zum Beweise leiten wir aus (5) zunächst ab:

$$(6) \qquad \int\limits_{-A}^{B} f_x(x, y)\,dy = \frac{1}{\pi}\int\limits_{-\frac{\pi}{2}}^{+\frac{\pi}{2}} d\varphi \int\limits_{-A}^{B} F_p(x\cos\varphi + y\sin\varphi, \varphi)\cos\varphi\,dy,$$

wo A, B zwei positive Konstante sind.

Wir setzen nun, wie schon früher analog geschehen ist:

$$F(p, \varphi) = F(p, 0) + \sin\varphi\, G(p, \varphi),$$

wobei $G(p, \varphi)$ im Integrationsgebiete beschränkt bleibt und für $p \to \infty$ den Grenzwert Null erhält. Aus:

$$\int\limits_{-A}^{B} G_p(x\cos\varphi + y\sin\varphi, \varphi)\cos\varphi \cdot \sin\varphi\,dy$$
$$= [G(x\cos\varphi + B\sin\varphi, \varphi) - G(x\cos\varphi - A\sin\varphi, \varphi)]\cos\varphi$$

folgt, daß der zweite Bestandteil von (6) für $A \to \infty$, $B \to \infty$ den Grenzwert Null erhält, so daß nur der erste zu untersuchen bleibt. Durch die analoge Integration erkennt man, daß in diesem ersten Bestandteil das Integral nach φ über jedes $\varphi = 0$ *nicht* enthaltende Intervall für $A \to \infty$, $B \to \infty$ ebenfalls Null wird; es bleibt also zu betrachten:

$$\lim_{\substack{A\to\infty \\ B\to\infty}} \frac{1}{\pi}\int\limits_{-\varepsilon}^{+\varepsilon} d\varphi \int\limits_{-A}^{B} F_p(x\cos\varphi + y\sin\varphi, 0)\cos\varphi\,dy, \qquad 0 < \varepsilon < \frac{\pi}{2}.$$

Man kann dieses Integral schreiben:

$$\frac{1}{\pi}\int\limits_{-\varepsilon}^{+\varepsilon} d\varphi \int\limits_{x\cos\varphi - A\sin\varphi}^{x\cos\varphi + B\sin\varphi} F_p(p, 0)\,\mathrm{ctg}\,\varphi\, dp$$

und erhält daraus, wenn A und B genügend groß angenommen werden, durch Vertauschung der Integrationsfolge nach einiger Rechnung den Wert:

$$\frac{1}{\pi}\int\limits_{x\cos\varepsilon - B\sin\varepsilon}^{x\cos\varepsilon + B\sin\varepsilon} l\frac{(B^2+x^2)\sin\varepsilon}{|Bp - x\sqrt{B^2+x^2-p^2}|}\, F_p(p, 0)\, dp$$

$$+\frac{1}{\pi}\int\limits_{x\cos\varepsilon - A\sin\varepsilon}^{x\cos\varepsilon + A\sin\varepsilon} l\frac{(A^2+x^2)\sin\varepsilon}{|Ap - x\sqrt{A^2+x^2-p^2}|}\, F_p(p, 0)\, dp.$$

Es genügt, den Grenzwert des zweiten Integrales für $A \to \infty$ zu ermitteln. Wir schreiben es so:

$$\frac{1}{\pi} l(A\sin\varepsilon)[F(x\cos\varepsilon + A\sin\varepsilon, 0) - F(x\cos\varepsilon - A\sin\varepsilon, 0)]$$

$$+\frac{1}{\pi}\int\limits_{x\cos\varepsilon - A\sin\varepsilon}^{x\cos\varepsilon + A\sin\varepsilon} l\frac{1}{|p-x|}\, F_p(p, 0)\, dp$$

$$+\frac{1}{\pi}\int\limits_{x\cos\varepsilon - A\sin\varepsilon}^{x\cos\varepsilon + A\sin\varepsilon} l\frac{|Ap + x\sqrt{A^2+x^2-p^2}|}{A|p+x|}\, F_p(p, 0)\, dp.$$

Da der Logarithmus in dem letzten Integral für $A \to \infty$ *gleichmäßig* gegen Null geht, so folgt als Grenzwert:

$$-\frac{1}{\pi}\int\limits_{-\infty}^{+\infty} F_p(p, 0)\, l\,|p-x|\, dp,$$

womit der Grenzwert von (6) erhalten wird:

$$\int\limits_{-\infty}^{+\infty} f_x(x, y)\, dy = -\frac{2}{\pi}\int\limits_{-\infty}^{+\infty} F_p(p, 0)\, l\,|p-x|\, dp.$$

Es mag hier bemerkt werden, daß letzterer Ausdruck die Randwerte des Imaginärteils einer in der oberen Halbebene regulären analytischen Funktion darstellt, für welche die Randwerte des Realteiles den Wert $2F(x, 0)$ haben.

Bilden wir jetzt im Sinne der Formel (V):

$$-\int_{-\infty}^{+\infty}\frac{dx}{x}\int_{-\infty}^{+\infty}f_x(x,y)\,dy=\frac{2}{\pi}\int_0^{\infty}\frac{dx}{x}\int_{-\infty}^{+\infty}F_p(p,0)\,l\left|\frac{p-x}{p+x}\right|dx,$$

so ist dieses Doppelintegral absolut konvergent und führt wegen

$$\int_0^{\infty}l\left|\frac{p-x}{p+x}\right|\frac{dx}{x}=-\frac{\pi^2}{2}\operatorname{sgn}p$$

genau zur Formel (V).

4. Es sei nun f eine Punktfunktion mit folgenden Regularitätseigenschaften:

a_4) f sei mit seinen Ableitungen bis zur zweiten Ordnung einschließlich stetig.

b_4) Die Ausdrücke

$$f(x,y),\quad \sqrt{x^2+y^2}\,l(x^2+y^2)f_x(x,y),\quad \sqrt{x^2+y^2}\,l(x^2+y^2)f_y(x,y)$$

haben für $x^2+y^2\longrightarrow\infty$ den Grenzwert Null.

c_4) Die Integrale

$$\int_{-\infty}^{+\infty}\int_{-\infty}^{+\infty}D_1f\cdot\frac{l(x^2+y^2)}{\sqrt{x^2+y^2}}\,dx\,dy\quad\text{und}\quad\int_{-\infty}^{+\infty}\int_{-\infty}^{+\infty}D_2f\cdot l(x^2+y^2)\,dx\,dy,$$

wo D_1f jede erste, D_2f jede zweite Ableitung von f bedeutet, konvergieren absolut.

Diese Bedingungen sind wieder gegenüber Bewegungen invariant.

Dann gilt:

Satz VI: Die aus f nach (V) gebildete Geradenfunktion besitzt die Punktmittelwerte $f(x,y)$.

Es genügt, den Beweis für den Nullpunkt zu erbringen. Für eine beliebige Gerade durch diesen ergibt (V) nach einer partiellen Integration:

$$F(0,\varphi)=\frac{1}{2\pi}\int_{-\infty}^{+\infty}\int_{-\infty}^{+\infty}[f_{xx}\cos^2\varphi+2f_{xy}\sin\varphi\cos\varphi$$

$$+f_{yy}\sin^2\varphi]\,l\,|x\cos\varphi+y\sin\varphi|\,dx\,dy$$

oder nach Einführung von Polarkoordinaten ϱ, ψ:

$$F(0, \varphi) = \frac{1}{2\pi}\int_0^\infty \varrho\, d\varrho \int_0^{2\pi}\Big[\frac{\partial^2 f}{\partial \varrho^2}\cos^2(\varphi-\psi)$$

$$+ 2\frac{\partial^2 f}{\partial\varrho\,\partial\psi}\frac{\sin(\varphi-\psi)\cos(\varphi-\psi)}{\varrho} + \frac{\partial^2 f}{\partial\psi^2}\frac{\sin^2(\varphi-\psi)}{\varrho^2}$$

$$+ \frac{\partial f}{\partial\varrho}\frac{\sin^2(\varphi-\psi)}{\varrho}$$

$$- 2\frac{\partial f}{\partial\psi}\frac{\sin(\varphi-\psi)\cos(\varphi-\psi)}{\varrho^2}\Big] l\,|\,\varrho\cos(\varphi-\psi)\,|\, d\psi.$$

Um den Punktmittelwert für $[0, 0]$ zu bilden, kann man die Integration nach φ unter dem Doppelintegral von 0 bis 2π ausführen und hat dann noch durch 2π zu dividieren. Das Glied mit $\frac{\partial^2 f}{\partial\psi^2}$, das dabei zum Vorschein kommt, fällt vermöge Integration nach ψ weg und es bleibt:

$$\frac{1}{2\pi}\int_0^{2\pi} d\psi \int_0^\infty \Big[\frac{1}{4}\Big(\varrho\frac{\partial^2 f}{\partial\varrho^2} - \frac{\partial f}{\partial\varrho}\Big) + \frac{1}{2}\,l\,\frac{\varrho}{2}\,\frac{\partial}{\partial\varrho}\Big(\varrho\frac{\partial f}{\partial\varrho}\Big)\Big] d\varrho,$$

was sich in der Tat auf $f(0, 0)$ reduziert.

Um die eindeutige Bestimmtheit von F zu zeigen, wären die Bedingungen a_3—c_3 als erfüllt nachzuweisen, was offenbar noch weitere Voraussetzungen über f nötig macht.

5. Es möge hier folgende Bemerkung Platz finden, die ich, wie überhaupt die Problemstellung B, Herrn W. Blaschke verdanke: Die beiden hier behandelten Aufgaben stehen in engem Zusammenhange mit der Theorie des Newtonschen Potentials. Betrachten wir nämlich den Übergang von einer Punktfunktion f zu ihren geradlinigen Mittelwerten F als eine lineare Funktionaltransformation

$$F = Rf,$$

und ebenso den Übergang von einer Geradenfunktion F zu ihren Punktmittelwerten v:

$$v = BF,$$

so liegt es nahe, die zusammengesetzte, durch

$$v = Hf = B[Rf] = BRf$$

definierte Transformation $H = BR$ zu betrachten.

Man sieht nun sofort, daß Hf nichts anderes ist, als das NEWTONsche Potential der mit der Massendichte $\frac{1}{\pi}f$ belegten Ebene in den Punkten der Ebene selbst. Daraus kann man nach einer Bemerkung von G. HERGLOTZ die Umkehrung der Transformation H gewinnen; es ergibt sich dabei:

$$f(P) = H^{-1}v = -\frac{1}{2}\int_0^\infty \frac{d\bar{v}_P(r)}{r} = -\frac{1}{4\pi}\int_{-\infty}^{+\infty}\int_{-\infty}^{+\infty}\frac{\Delta v(x'y')}{r_{PP'}}dx'dy',$$

wo $\bar{v}_P$ eine zu den früher eingeführten analoge Mittelwertsbezeichnung ist und Δ den LAPLACEschen Operator bedeutet.

Nun liegt der Gedanke nahe, die in 1.—3. direkt geleitete Umkehrung von R und H durch den Ansatz

$$R^{-1} = H^{-1}B \quad \text{bzw.} \quad B^{-1} = RH^{-1}$$

zu leisten. Tatsächlich habe ich die Umkehrungsformel (IV) zuerst auf diesem Wege gefunden, aber eine strenge Durchführung dieses Gedankens scheint schwieriger zu sein als die direkte Verifikation und versagt auch in den gleich zu besprechenden nichteuklidischen Fällen.

Schließlich sei bemerkt, daß die in A und B zugrunde gelegten Regularitätsbedingungen selbstredend bei weitem nicht die allgemeinsten sind, wie sich an einfachen Beispielen zeigen läßt.

C. Verallgemeinerungen.

6. Eine weitgehende Verallgemeinerung des in A behandelten Problems ließe sich etwa folgendermaßen formulieren: es sei eine Fläche S gegeben, auf der irgendwie ein Bogendifferential ds definiert sei, ferner eine zweifach unendliche Schar von Kurven C auf S. Es soll eine Punktfunktion der Fläche aus ihren Integralwerten $\int f ds$ längs der Kurven C bestimmt werden.

Die nächstliegende Spezialisierung erhält man, wenn man für S eine nichteuklidische Ebene, für ds das zugehörige Bogenelement und für die Kurven C die Geraden nimmt. Im elliptischen Falle kann man die Aufgabe auf die Kugelgeometrie hinüberspielen; indem man in bekannter Weise ein diametrales Punktepaar der Kugel als Punkt der elliptischen Ebene deutet, erhält man die Aufgabe, eine gerade — d. h. in diametralen Punkten gleichwertige — Funktion auf der Kugel aus ihren Großkreisintegralen zu bestimmen. Minkowski hat diese Aufgabe im Prinzip zuerst behandelt[1]) und durch Entwicklung nach Kugelfunktionen gelöst; P. Funk hat später die Minkowskische Lösung durchgeführt und gezeigt, wie man die Lösung mit Hilfe der Abelschen Integralgleichung finden kann[2]) und dieser Methode verdanke auch ich die Lösung des Problems A. Die Funksche Lösung ist ganz analog zu (III), nur tritt in den Nenner der Sinus des sphärischen Radius und additiv zu dem Integral der Wert von F im Pole des betreffenden Großkreises dividiert durch π. Aber auch in der hyperbolischen Ebene hat die gestellte Aufgabe die zu (III) analoge Lösung:

$$f(P) = -\frac{1}{\pi}\int_0^\infty \frac{dF_P(q)}{\mathfrak{Sin}\, q}$$

(es ist hier das Krümmungsmaß $= -1$ angenommen), wie sich ganz konform zu der in 1. angedeuteten Ableitung von (III) zeigen läßt.

In beiden Fällen kann man auch die zu B. analoge Frage stellen. In der elliptischen Geometrie erhält man vermöge der absoluten Polarität nichts Neues, im hyperbolischen Falle scheint eine zu (V) analoge Lösung nicht zu existieren.

Eine zweite Spezialisierung ergibt sich, wenn man (in der euklidischen oder nichteuklidischen Geometrie) als Kurven C die Kreise mit konstantem Radius nimmt. Hier kann man auf der Kugel die Minkowskische Behandlung mit Kugelfunktionen anwenden und die Aufgabe in gewissem Grade lösen. Interessant ist in diesem Falle, daß die Eindeutigkeit der Lösung verloren gehen kann; es gibt nämlich für gewisse, durch die Nullstellen der Legendreschen Polynome gerader Ordnung definierte Radien ϱ gerade Funktionen auf der Kugel, die längs jedes Kreises vom sphärischen Radius ϱ integriert, Null ergeben, ohne identisch zu verschwinden. Im euklidischen Falle tritt an Stelle der Kugelfunktionenreihen das Integraltheorem der Besselfunktionen; hier gibt es stets Funktionen, die über alle Kreise von festem Radius integriert, Null geben und doch nicht identisch verschwinden; ist der Radius 1, so sind es (in Polarkoordinaten ϱ, φ) die Funktionen

$$J_n(x_\nu \varrho)\cos n\varphi, \quad J_n(x_\nu \varrho)\sin n\varphi,$$

und ihre Linearkombinationen, wo x_ν eine Nullstelle von J_0 ist. Im hyperbolischen Falle treten an Stelle der Besselfunktionen die

1) Ges. Abh., Bd. II, S. 277 f. 2) Math. Ann., Bd. 74, S. 283—288.

sogenannten Kegelfunktionen, für die das zugehörige Integraltheorem von WEYL[1]) bewiesen ist. Die Ergebnisse sind analog zum euklidischen Falle.

7. In anderer Richtung verallgemeinern sich die Ergebnisse von A und B beim Übergang zu höheren Räumen. In einem euklidischen R_n kann man eine Punktfunktion $f(P) = f(x_1, x_2, \ldots x_n)$ aus ihren Integralwerten $F(\alpha_1 \ldots \alpha_n, p)$ über alle Hyperebenen $\alpha_1 x_1 + \cdots + \alpha_n x_n = p$ $(\alpha_1^2 + \cdots + \alpha_n^2 = 1)$ zu bestimmen trachten. Analog zu dem in 1. eingehaltenen Vorgang bilden wir den Mittelwert $\bar{F}_0(q)$ von F über die Tangentialebenen der Kugel vom Zentrum $[0, 0 \ldots 0]$ und Radius q. Er ist durch das $n - 1$fache Integral:

$$\bar{F}_0(q) = \frac{1}{\Omega_n} \int F(\alpha, q)\, d\omega$$

gegeben, wo $d\omega$ das Oberflächenelement, $\Omega_n = \dfrac{2\pi^{\frac{n}{2}}}{\Gamma\left(\frac{n}{2}\right)}$ die Oberfläche der n-dimensionalen Kugel $\alpha_1^2 + \cdots + \alpha_n^2 = 1$ bedeutet.

Man kann $\bar{F}_0$ durch ein n-faches Integral über f darstellen und zwar ergibt sich:

$$(7) \quad \bar{F}_0(q) = \frac{\Omega_{n-1}}{\Omega_n} \int \cdots \int_{x_1^2 + \cdots + x_n^2 > q^2} f(x_1, x_2, \ldots x_n) \frac{(x_1^2 + \cdots + x_n^2 - q^2)^{\frac{n-3}{2}}}{(x_1^2 + \cdots + x_n^2)^{\frac{n-2}{2}}}\, dx_1 \ldots$$

oder in einer nun schon oft verwendeten Mittelwertsbezeichnung:

$$\bar{F}_0(q) = \Omega_{n-1} \int_q^\infty f_0(r)(r^2 - q^2)^{\frac{n-3}{2}}\, r\, dr.$$

Dies ist die zu (1) analoge Formel, an die sich entsprechende Folgerungen anschließen. Die Substitution $r^2 = v$, $q^2 = u$ führt auf die Integralgleichung:

$$\Phi(u) = \frac{\Omega_{n-1}}{2} \int_u^\infty \varphi(v)(v - u)^{\frac{n-3}{2}}\, dv.$$

---

1) Gött. Nachr. 1910, S. 454.

Ist n *gerade*, so ergibt $\left(\frac{n}{2}-1\right)$-maliges Differentiieren nach n dieselbe Gleichung wie (2) und man kann hieraus

$$\varphi(0) = f(0, 0, \ldots 0)$$

finden, wozu also bei gegebenem F die Bildung von F, Differentiationen und eine Integraloperation nötig sind. Bei *ungeradem* n fällt diese Integraloperation weg, denn jetzt ergibt $\left(\frac{n-1}{2}\right)$-maliges Differentiieren:

$$\varphi(0) = \frac{2(-1)^{\frac{n-1}{2}}}{\Omega_{n-1}\left(\frac{n-3}{2}\right)!}\,\Phi^{\left(\frac{n-3}{2}\right)}(0).$$

Besonders einfach gestaltet sich der dreidimensionale Fall; diesen kann man aber auch nach einer Methode behandeln, die zu 5. analog ist und sehr elegante Ergebnisse liefert. Aus (7) geht nämlich für $q = 0$ der Punktmittelwert von F hervor:

$$F_0 = \frac{1}{2}\iiint \frac{f(x, y, z)}{\sqrt{x^2 + y^2 + z^2}}\,dx\,dy\,dz,$$

der als das Newtonsche Potential des mit der Massendichte $\frac{1}{2}f$ belegten Raumes anzusprechen ist. Also folgt:

$$f(x, y, z) = -\frac{1}{2\pi}\Delta F,$$

wo F den Punktmittelwert von F bedeutet.

Hier kann man auch die zu B. analoge Aufgabe lösen und findet nach der in 5. angedeuteten Methode für eine Ebenenfunktion F, deren Punktmittelwerte f bekannt sind:

$$F(E) = -\frac{1}{2\pi}\iint \Delta f\,d\sigma,$$

wo $d\sigma$ das Flächenelement der Ebene E ist. Δ ist der Laplacesche Operator für den dreidimensionalen Raum, die Integration über die ganze Ebene E zu erstrecken.

Druckfertig erklärt 1. X. 1917.]